STUDIES IN CANADIAN REGIONAL GEOGRAPHY

Essays in Honor of J. Lewis Robinson

B.C. GEOGRAPHICAL SERIES, NUMBER 37

Edited by

Brenton M. Barr
Professor of Geography
The University of Calgary

Tantalus Research Limited
Publisher - Vancouver, Canada

B.C. GEOGRAPHICAL SERIES, NUMBER 37

Editorial Address:
W.G. Hardwick, Ph.D.
Department of Geography
The University of British Columbia
2075 Wesbrook Place
Vancouver, B.C. V6T 1W5

Circulation Address:
B.C. Geographical Series
P.O. Box 34248
2405 Pine Street
Vancouver, B.C. V6J 4N8

ISBN 0-919478-58-1

Canadian Cataloguing in Publication Data

Main entry under title:

Studies in Canadian regional geography

(B.C. geographical series, ISSN 0068-1571 ; no. 37)
ISBN 0-919478-58-1

1. Canada - Decription and travel - 1950- Addresses, essays, lectures.* I. Barr, Brenton M., 1940- II. Robinson, J. Lewis (John Lewis), 1918- III. Series.
FC75.S88 1984 917.1 C84-091146-7
F1016.S88 1984

PRINTED IN CANADA
by
G.A. Roedde Ltd.

CONTENTS

EDITOR'S PREFACE

The editor wishes to thank all those who contributed essays to this volume and otherwise encouraged its preparation. Special thanks are due to Professor J. Lewis Robinson who graciously consented to provide photographs, career and bibliographical data, and valuable insights throughout much of 1983 despite the heavy demands placed on his time by numerous matters associated with his impending retirement. Professor John K. Stager, Associate Dean of Arts at the University of British Columbia provided helpful suggestions and important perspectives. Professor David Hardwick, a principal of Tantalus Research Limited and established friend of Canadian geography, offered enthusiastic support for publication of these essays. Mr. Robert Martin, Director of Research Services at the University of Calgary, provided financial assistance without which the volume could not have been published. Ms. Doris Carlson of the Geography Department, University of Calgary, cheerfully typed the manuscripts in camera-ready form despite many demands placed on her time by other administrative requests by the editor and members of the department. The editor regrets that, due to limitations of space and resources, all those wishing to participate in this volume could not be accommodated.

Brenton M. Barr,
Calgary, January 1984

Head, Department of Geography, U.B.C., 1963

SELECT ACADEMIC PROFILE OF J. LEWIS ROBINSON

Biography and Career

John Lewis Robinson, born 9 July 1918, in Leamington, Ontario

Graduated from University of Western Ontario, 1940, B.A. (Hons. History)
Exchange scholarship student to University of British Columbia, 1938-39

Graduated from Syracuse University, New York, 1942, M.A. (Geography)
Thesis: *Urban Geography of Windsor, Ontario*

Graduated from Clark University, Worcester, Mass., 1946, Ph.D. (Geography) Dissertation: *The Canadian Eastern Arctic, A Regional Geography*

Geographer, Northwest Territories Administration, Government of Canada, Department of Mines and Resources, 1943-46

Appointed to the University of British Columbia, 1946

Chairman, Geography Division, Department of Geology and Geography, University of British Columbia, 1953-59

Head, Department of Geography, University of British Columbia, 1959-68

Acting Head, Department of Geography, University of British Columbia, 1974-75

Retired as Professor of Geography, University of British Columbia, 31 December, 1983

Professional Awards, Distinctions

Vice-President, Association of Pacific Coast Geographers, 1949-50

President, Association of Pacific Coast Geographers, 1950-51

Vice-President, Canadian Association of Geographers, 1951-52

President, Canadian Association of Geographers, 1955-56

Chairman, British Columbia Division, Canadian Association of Geographers, 1958

British Council Travelling Fellowship to departments of geography in Britain, 1960

Canada Council Senior Fellowship, 1965-66

Citation of Merit, Association of American Geographers, 1966

Centennial Medal, Government of Canada, 1967

Massey Medal, Royal Canadian Geographical Society, 1971

Elected Fellow, Geographical Society of Chicago, 1973

Canada Council, Leave Fellowship, 1973-74

Honorary Life Member, University of British Columbia Alumni Association, 1975

Award for Service to the Profession of Geography, Canadian Association of Geographers, 1976

University of British Columbia Certificate of Merit for Outstanding Teaching, 1976

University of British Columbia Master Teacher Award, 1977

Canadian Silver Jubilee Medal, Government of Canada, 1977

Distinguished Teaching Award for Canada, National Council for Geographic Education (U.S.A.), 1982

Graduation from University of Western Ontario, 1940

Graduate Student in Geography at Syracuse University, New York, 1941

PUBLICATIONS OF J. LEWIS ROBINSON

Geography as a Discipline in Canada

(1950) The Geographical Basis of Foreign Trade, *Canadian Geographical Journal*, (March): 114-123.

(1950) The Use of Maps, in *Social Studies Curriculum Guide* (Victoria: Department of Education): 29-30.

(1951) The Development and Status of Geography in the Universities and Government in Canada (Presidential Address), *Yearbook of the Pacific Coast Geographers*, 13: 3-13.

(1951) Geography in the Universities of Canada, *Canadian Geographical Journal*, (October): 180-183.

(1952) *The Status of Geography in Countries Adhering to IGU--Canada* (Washington: International Geographical Union, Publication 7): 10-13.

(1954) The Place of Geography in the Social Studies, *B.C. Teacher*, 33: 160-163. (Reprinted in Denoyer-Geppert *Memo Book for Teachers*, 1955; and in *Towards Better Understanding and Use of Maps* (Chicago: Denoyer-Geppert Map Co., 1957): 1-2)

(1955) Textbooks on the Geography of Canada, review article in *Canadian Geographical Journal*, (February): v-xiii.

(1956) The Problems of Geographical Regions in Canada, *The Canadian Geographer*, 7: 46-49.

(1956) Geography and Regional Planning (Presidential Address), *The Canadian Geographer*, 8: 1-8.

(1956) The Canadian Association of Geographers, *The Professional Geographer*, 8(May): 9-10.

(1957) 18th Congress of the International Geographical Union (in Brazil), *Canadian Geographical Journal*, (March): 118-124.

(1959) Geography at the University of British Columbia, *The Canadian Geographer*, 13: 46-47.

(1961) Applied Geography, *Report of Canadian Committee of IGU* (Ottawa): 52-56.

(1966) The Need for Urban Geography in our High Schools, *Journal of Geography*, (May): 236-240.

(1967) Growth and Trends in Geography in Canadian Universities, *The Canadian Geographer*, 11(4): 216-229.

(1969) Trends in Geography in Canadian Universities, *B.C. Geographical Series*, (8): 47-56. (Summarized in the *Journal of Geography*, (April, 1969): 207)

(1974) Geography of Canada Courses in Canadian Universities, 1973-74, in *Newsletter of the Canadian Association of Geographers:* ix-xiv, and in *Newsletter of the Canadian Committee of IGU*, (2, 1974): 10-13.

(1975) Comment on the Locative Function of Country Names, *The Canadian Geographer*, 19(4): 347-48.

(1976) A New Look at the Four Traditions of Geography, *Journal of Geography*, 75(9): 520-530.

(1977) The Production and Employment of Geographers in Canada, *The Professional Geographer*, 29(2): 208-214.

(1977) Comment on the Geography Section of the Symons Report on Canadian Studies, *University Affairs*, (May): 10.

(1978) Ph. D. Geographers in Canada, 1977-78: Supply and Demand, in *Newsletter of the Canadian Committee for Geography*, (April): 8-13.

(1979) Geography in Canada 25 Years Ago, in *Retrospective 1951-76* (Montreal: Canadian Association of Geographers): 19-30.

(1979) Geography, in Discipline Reviews of the Symons Report, *Canadian Issues*, 2(1): 115-116.

(1982) Mental maps in the Classroom, *Horizon*, 20(January): 18-20. (Reprinted with minor changes in The *Monograph*, 33(2, 1982): 8-9)

(1983) *Careers in Geography* (Montreal: The Canadian Association of Geographers), six-panel leaflet.

(1984) The Employment of Ph.D. Geographers in Canada, *The Operational Geographer*, (3) (forthcoming).

(1984) Geography in Canada, in *Encyclopedia Canadiana* (Edmonton), (forthcoming).

(1984) Human Geography in Canada, in *Encyclopedia Canadiana* (Edmonton), (forthcoming).

Geography of Canada (excluding publications listed in other sections)

(1950) *The Geography of Canada* (Toronto: Longmans Green): 200 pp. (Reprinted by Greenwood Press, Connecticut, 1970)

(1951) Climate and Man in Canada, *Proceedings of 19th Western Snow Conference* (Fort Collins, Colorado): 1-12.

(1952) *Physical-Political Wall Map of Canada* (Chicago: Denoyer-Geppert Map Co.), Scale 50 miles: 1 inch. (Revised editions: 1956, 1959, 1962, 1967, 1970)

(1954) Northward by Road and Rail (in Canada), *The Beaver Magazine*, (June): 43-45.

(1957) Canada and the North American Arctic, in *World Political Geography* (New York: Crowell): 77-92.

(1960) Canada, Alaska and Greenland, in *World Geography* (New York: McGraw-Hill): 112-149. (Revised 1960)

(1965) Canada and Greenland, in J. Morris (editor), *World Geography*. Second edition (New York: McGraw-Hill: 132-171. (Rewritten for Third edition, (1972): 102-137)

(1958) Kanada, *Verlag Herder Atlas and Encyclopedia* (Breisgau, West Germany): 508-413.(Translated into German)

(1959) Canada, in *Encyclopedia Britannica*, 4: 690-714. (Canada: Resources, in 1965 edition; Physical Geography, in 1971 edition, 4: 726-729)

(1959) Canada, in *Encyclopedia Britannica Junior*, 4: 50-72. (In 1971 edition, 4: 50-62L)

(1962) Research Organizations in Canada, *Proceedings of Resources for Tomorrow Conference* (Ottawa): 3: 252-254.

(1963) (editor) Canada, in *Illustrated Atlas of Today's World* (Chicago: Rand McNally Co.), 6: 454-539.

(1964) New Industrial Towns on Canada's Resource Frontier, review article in *The Geographical Review*, 54(2): 289-291.

(1964) *Four Wall Maps of Canada (Population, Land Use, Manufacturing, Political)*, Canadian History Series (Chicago: Denoyer-Geppert), Numbers 13-16. (*Manufacturing and Mineral* Map Revised, 1973)

(1965) Canada, *Chambers Enclyclopedia* (London, England), 3: 2-18.

(1966) North America Textbooks, review article in *The Geographical Review,* 56(2): 301-303.

(1966) Geographical Regions of Canada, *Proceedings of 18th Congress of International Geographical Union, Rio de Janeiro, 1956,* 4: 654-658.

(1969) (with O.P. Starkey) *The Anglo-American Realm* (New York: McGraw-Hill). Authorof four regional chapters on Canada: 404-496.

(1969) *Resources of the Canadian Shield* (Toronto: Methuen): 136 pp.

(1971) *Political Wall Map of Canada* (with shading for settlement) (Chicago: Denoyer-Geppert Map Co.), Scale 64 miles: 1 inch.

(1972) (with George Tomkins) (co-editor) Canada, in *Gage World Atlas: A Canadian Perspective* (Toronto: Gage Educational Publishers): 11-49.

(1972) Regional Geography of Canada, in *1972 Canada Year Book* (Ottawa: Statistics Canada): 1-26. (Reprinted for distribution to International Geographical Union Montreal Congress)

(1973) *Shaded Relief Wall Map of Canada* (Chicago: Denoyer-Geppert Map Co.), Scale 50 miles: 1 inch.

(1975) (with O.P. Starkey and C. Miller) (co-editor) *The Anglo-American Realm.* Second edition (New York: McGraw-Hill). Author of chapters on Canada: 267-340.

(1977) *Canada. Physical-political Wall Map.* Simplified edition (Chicago: Denoyer-Geppert Map Co.).

(1979) *Regional Geography of Canada. Correspondence courses Geography 470 and 471* (Richmond, B.C.: Open Learning Institute): 300 pp.

(1980) Mental Maps of Canada, *Journal of Geography,* 79(4): 124-125.

(1981) The Varied Mental Maps (of Canada) our Students have, *Canadian Geographic,* (April-May): 52-57.

(1981) Regional Geography of Canada, *Canada Handbook* (Ottawa: Statistics Canada): 1-17.

(1981) Regional Geography of Canada in Canada, *Journal of Geography,* 80(7): 268-271.

(1982) *The Physical Environment of Canada and the Evolution of Settlement Patterns* (Vancouver: Talonbooks): 43 pp.

(1983) *Concepts and Themes in the Regional Geography of Canada* (Vancouver: Talonbooks): 345 pp.

Northern Canada

(1944) *An Outline of the Canadian Eastern Arctic--Its Geography, Peoples and Problems* (Ottawa: Department of Mines and Resources, Bureau of Northwest Territories): 38 pp.

(1944) Eskimo Population in the Canadian Eastern Arctic, *Canadian Geographical Journal*, (September): 128-142.

(1944) Mineral Resources and Mining Activity in the Canadian Eastern Arctic, *Canadian Geographical Journal*, (August): 55-75.

(1944) The Battle of Fort Ross (Ice Navigation in the Arctic Islands), *The Beaver Magazine*, (March): 4-7.

(1945) Conquest of the Northwest Passage by the RCMP Schooner St. Roch, *Canadian Geographical Journal*, (February): 52-73. (Reprinted in the *Annual Report of the Smithsonian Institute*, Washington, D.C., 1945: 219-234; condensed and reprinted in *Patterns of Canada*, Canadian Geographical Society, Ottawa, 1966 (Part 4): 152-159)

(1945) (with P.D. Baird) A Brief History of Exploration and Research in the Canadian Eastern Arctic, *Canadian Geographical Journal*, (March): 136-157.

(1945) Land Use Possibilities in Mackenzie District, N.W.T., *Canadian Geographical Journal*, (July): 30-47.

(1945) Agriculture and Forests of Yukon Territory, *Canadian Geographical Journal*, (August): 54-72.

(1945) Water Transportation in the Canadian Northwest, *Canadian Geographical Journal* (November): 236-256.

(1945) Physical Geography of the Canadian Eastern Arctic, *Canada Year Book*: 12-19.

(1946) (with M.J. Robinson) Fur Production in the Northwest Territories, *Canadian Geographical Journal*, (January): 34-48. (Reprinted in *The New Northwest*, University of Toronto Press, 1947: 133-156)

(1946) Weather and Climate of the Northwest Territories, *Canadian Geographical Journal*, (March): 124-139.

(1946) (with M.J. Robinson) Exploration and Settlement of Mackenzie District, N.W.T., *Canadian Geographical Journal*, Part 1 (June): 246-255; Part 2 (July): 43-49.

(1946) The Canadian Northland, *The Canadian Banker*, 53: 84-94.

(1947) RMS Nascopie-Veteran of the Seas, *Canadian Geographical Journal*, (October): 196-197.

(1947) Polar Regions, in *Colliers Encyclopedia*, 16: 159-162. (Revised for 17th edition, 1958: 159-162)

(1948) Canada's Western Arctic, *Canadian Geographical Journal*, (December): 242-260.

(1948) *Physical Geography of the Canadian Western Arctic*, Canada Year Book: 9-18.

(1949) Resources of the Canadian Arctic, *The Beaver Magazine*, (December): 48-51.

(1949) Problems of Arctic Geography in Canada, abstract in *Annals of the Association of American Geographers*, (March): 75.

(1951) Changing the Arctic Maps, *The Beaver Magazine*, Part 1 (December): 8-9: Part 2 (March, 1952): 24-26.

(1952) *The Canadian Arctic, Information Bulletin No. 2* (Ottawa: Geographical Branch): 118 pp.

(1952) The Canadian Northland, in D.F. Putnam (editor), *Canadian Regions* (Toronto: J.M. Dent): 480-508. (Revised 1956)

(1954) Climate of Northern Canada, in C. Wilson (editor), *North of 55°* (Toronto: Ryerson Press): 32-43.

(1954) Mapping in the North American Arctic, review article in *The Geographical Review*, 44(2): 290-292.

(1959) Arctic Resources, *The Beaver Magazine*, (Spring): 9-11.

(1959) Dawson, in *Encyclopedia Britannica*, 7: 91. (Revised 1967: 110)

(1959) Yukon River, in *Encyclopedia Britannica*, 23: 924-925 (Revised 1967: 931-932; expanded and rewritten 1973: 1107-1108)

(1959) Yukon Territory, in *Encyclopedia Britannica*, 23: 925-926. (Revised 1967: 932-934)

(1962) Northwest Territories,*American Peoples Encyclopedia* (Chicago), 14: 7097-7099. (Reprinted 1965, 14: 36-38)

(1962) Yukon Territory, *American Peoples Encyclopedia* (Chicago), 19: 10037-10038. (Reprinted 1965, 19: 464-465)

(1965) Northwest Territories, *Chambers Encyclopedia* (England), 10: 83.

(1965) Yukon Territory, *Chambers Encyclopedia* (England), 14: 792.

(1968) Northwest Territories, in *Crowell-Collier Merit Students Encyclopedia*, 13: 506-512.

(1968) Regional Geography of Hudson Bay Area, in *Science, History and Hudson Bay* (Ottawa: Department of Energy, Mines and Resources), Part 1: 201-235.

(1969) Great Slave Lake, Great Bear Lake, Back River, Mount Logan, short notes for *Colliers Merit Student Encyclopedia*.

(1973) Mackenzie River, in *Encyclopedia Britannica*, 14: 265-267.

(1979) Baffin Bay, Baffin Island, Banks Island, Beaufort Sea, short notes for *Funk and Wagnalls Encyclopedia*.

(1979) Search for the North Magnetic Pole, *The Beaver Magazine*, (Spring): 12-19.

(1980) Agriculture in the Territories of Northwestern Canada, in *Proceedings of the Conference on Boreal Agriculture* (Vancouver: University of British Columbia, Faculty of Agriculture): 1-7.

(1984) The Physical Environment of the Northwest Territories, *Alaska Geographic*, (forthcoming).

British Columbia

(1952) Dwindling Arable Lands in British Columbia, *Proceedings of 4th Regional Conference of B.C. Division, Community Planning Association* (Vancouver): 15-20.

(1953) Planning for Agriculture and Settlement in British Columbia, *Proceedings of the 6th Regional Conference of B.C. Division, Community Planning Association* (Vancouver): 10-13.

(1952) *Physical-Political Wall Map of British Columbia* (Chicago: Denoyer-Geppert Map Co.), Scale 20 miles: 1 inch.

(1953) The Rocky Mountain Trench, B.C., *The Beaver Magazine*, (March): 36-39.

(1953) Agricultural Regions of B.C., abstract in *Yearbook of Pacific Coast Geographers*, 15: 45-46.

(1954) (with J.D. Chapman) *British Columbia Geography Manual* (Victoria: B.C. Department of Education): 119 pp.

(1954) Population Trends and Distribution in B.C., *The Canadian Geographer*, 4: 27-32.

(1954) Population in B.C.--Trends, Densities and Distribution, *Proceedings of 7th B.C. Natural Resources Conference*, (Victoria): 210-218.

(1955) Map of B.C. Population Distribution, in *B.C. Atlas of Resources* (Victoria), Map. No. 16: 34.

(1958) British Columbia, in *Colliers Encyclopedia*, 4: 103-114.

(1958) The B.C. Natural Resources Conference, review article in *The Geographical Review*, 48(2): 300-301.

(1959) British Columbia, in *Funk and Wagnalls Encyclopedia*, 4: 1455-1470.

(1959) Strait of Georgia, in *Encyclopedia Britannica*, 10: 203. (Revised 1967, 10: 226)

(1959) Haro Strait, in *Encyclopedia Britannica*, 11: 215. (Revised 1967, 10: 121)

(1959) Juan de Fuca Strait, in *Encyclopedia Britannica*, 13: 160. (Revised 1967, 10: 98)

(1959) Kootenay River, in *Encyclopedia Britannica*, 13: 478. (Revised 1967, 10: 452)

(1960) What is Northern British Columbia?, *Proceedings of 12th B.C. Natural Resources Conference* (Victoria): 6-16.

(1960) British Columbia, in *World Book Encyclopedia*, 2: 504-517. (Revised 1965, 2: 504-517; partial revision 1969)

(1962) British Columbia, in *American Peoples Encyclopedia*, 4: 1590-1596. (1965, 3: 452-458; revised 1968)

(1965) British Columbia, in *Chambers Encyclopedia* (London, England), 2: 561-562.

(1965) Vancouver, in *Chambers Encyclopedia* (London, England), 14: 225-226.

(1965) Vancouver Island, in *Chambers Encyclopedia* (London, England), 14: 226.

(1965) Victoria, in *Chambers Encyclopedia* (London, England), 14: 296.

(1965) Nanaimo , B.C., *Canadian Geographical Journal*, (May): 162-69.

(1966) British Columbia, in *Grolier Book of Knowledge* (New York), 2: 402-407. (Revised 1971)

(1968) (with W.G. Hardwick) The Canadian Cordillera, in *Canada: A Geographic Interpretation* (Toronto: Methuen): 438-472.

(1968) The Rocky Mountain Trench of Eastern B.C., *Canadian Geographical Journal*, (October): 132-141.

(1970) Columbie Britannique, *Encyclopedia Universalis* (Paris), 4: 700-703. (Translated into French)

(1970) Vancouver, *Encyclopedia Universalis* (Paris), 4: 600.

(1972) (editor) *British Columbia, Studies in Canadian Geography* (Toronto: University of Toronto Press): 139 pp.

(1972) Areal Patterns and Regional Character, in *British Columbia, Studies in Canadian Geography* (Toronto: University of Toronto Press): 1-8.

(1973) (with W.G. Hardwick) *British Columbia - A Hundred Years of Geographical Change* (Vancouver: Talon Books): 63 pp.

(1973) Fraser River, in *Encyclopedia Britannica*, 9: 696-697.

(1974) How Vancouver has grown and changed, *Canadian Geographical Journal*, 89(4): 40-48.

(1975) *British Columbia, Physical-political Wall Map* (Chicago: Denoyer-Geppert Map Co.), Scale 20 miles: 1 inch.

(1980) British Columbia, in *Arete Encyclopedia* (Princeton, N.J.), 3: 489-492.

(1980) British Columbia, in *Grolier Book of Knowledge Encyclopedia* (Connecticut), 2: 402-407.

(1980) Vancouver, in *Grolier Book of Knowledge Encyclopedia* (Connecticut), 19: 276.

(1984) British Columbia, in *Encyclopedia Canadiana* (Edmonton) (forthcoming)

Prairie Provinces

(1957) The Northern Extension of the Pioneer Fringe of Agriculture on the Great Plains of Canada, *Proceedings of 17th Congress of International Geographical Union, 1952* (Washington, D.C.: U.S. Academy of Sciences, NRC): 657-662.

(1962) Saskatchewan, in *American Peoples Encyclopedia*, 16: 8417-8422. (Revised 1965, 16: 332-337; revised 1970)

(1965) Alberta, in *Chambers Encyclopedia* (England), 1: 229-230.

(1967) Milk River, in *Encyclopedia Britannica*, 15.

(1972) Lake Athabasca, Churchill River, Lake Louise, Nelson River, Reindeer Lake, Saskatchewan River, Lake Winnipeg, brief notes on lakes and rivers of Western Canada for *Crowell-Collier Encyclopedia*.

Ontario

(1943) Windsor, Ontario; a Study in Urban Geography, *Canadian Geographical Journal*, (September): 106-121.

(1962) Ontario, in *American Peoples Encyclopedia*, 14: 7214-7217. (Revised 1965, 14: 153-156; revised 1971)

Atlantic Provinces

(1962) New Brunswick, in *American Peoples Encyclopedia*, 14: 6928-6929. (Revised 1965, 13: 379-380)

(1962) Newfoundland, in *American Peoples Encyclopedia*, 14: 6935-6938. (Revised 1965, 13: 388-391)

(1962) Nova Scotia, in *American Peoples Encyclopedia*, 15: 7117-7119. (Revised 1965, 14: 56-58; revised 1968)

(1962) Prince Edward Island in *American Peoples Encyclopedia*, 15: 7827-7829. (Revised 1965, 14: 254-256; revised 1970)

(1965) Cape Breton, short note in *Chambers Encyclopedia* (England), 3: 68.

(1966) Maritime Provinces, in *Grolier Book of Knowledge* (New York), 12: 92-100.

(1966) Newfoundland, in *Grolier Book of Knowledge* (New York), 13: 141-145.

(1975) Changing Settlement Patterns in Newfoundland, review article in *The Geographical Review*, 65(1): 267-268.

Ice-bound with the R.M.S. Nascopie in the Central Arctic Islands, 1943

With Tom Weir (right), October, 1946. "The Beginnings of Geography at U.B.C."

INTRODUCTION

J. LEWIS ROBINSON AND CANADIAN REGIONAL GEOGRAPHY: AN ACKNOWLEDGMENT

Brenton M. Barr

The University of Calgary

This volume of essays has been prepared in honor of Professor J. Lewis Robinson to acknowledge the role which he has played in the lives and professional development of the contributors, all of whom are now engaged in the practice of geography in Canadian public institutions. All of the authors, whether engaged in research, planning, teaching, or various combinations of these pursuits, are dealing with regions and places, subjects and issues, whose intellectual antecedents were explored in the writing and teaching of J. Lewis Robinson. The volume includes geographers active in one or more of the regions (Figure 1) included in Robinson's analyses and whose specialization corresponds to at least one of the foci of Robinson's own work: northern Canada, Canadian regional geography, and conceptual representation of themes, problems and places pertaining to Canada's geography. Although diverse in their backgrounds and mentors, the authors in this volume are united in their commitment to the practice of geography, to the amplification of knowledge and understanding of spatial processes in Canada, and to the further development of a professional discipline whose role in Canada's scientific and educational development since World War II has been greatly enhanced by the work, inspiration and vision of Professor J. Lewis Robinson.

Organization of this Festschrift

To integrate the subjects addressed by each chapter, the contributors have ensured that their essays address subjects in at least one of the regions found in the writings of J. Lewis Robinson, and that their essays reflect changes in the foci of their subjects occurring in the period covered by Robinson's formal academic career, 1946-1984. The first three presentations, relating to northern Canada (Gilbert, Woo, and Bone), are followed by four chapters pertaining to aspects of applied

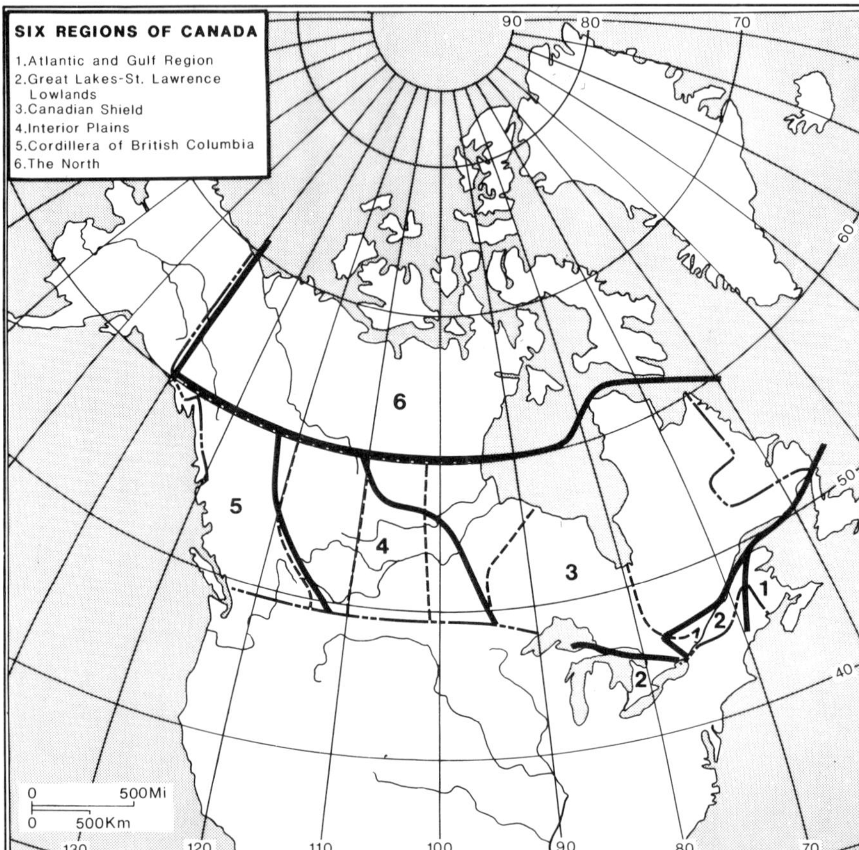

Figure 1. J. Lewis Robinson's Canadian Regions

Canadian geography in various regions of the country (Mitchell, Nicol, McAfee, and Gayler); the essay section of the volume concludes with three chapters probing different ways in which Canada's regional geography is perceived and analyzed (Young, Gibson, and Forward). The three essay sections correspond approximately to the major foci of J. Lewis Robinson's teaching and writing, and are intended to leave the reader with an awareness of the richness of insight and expertise present in the Canadian geographical community today. Prowess in geography could not have developed indigenously in Canada without the academic inspiration and creativity inspired in those attending Canadian universities since World War II by such leaders as J. Lewis Robinson.

Festschrifts for prominent geographers are important components of the literature in countries such as the United Kingdom, Australia and the United States, and hopefully will become more numerous in Canada as more founders of Canadian departments of geography and important contributors to geography as a discipline reach the age of retirement. The present volume seeks to ensure that the role of J. Lewis Robinson in Canada is acknowledged in a manner consistent with recognition of those outstanding members of the geographical community in other countries whose influence has spread throughout national domains and well into the international spheres of our profession. Despite variation in the type of sponsorship of the publications honoring outstanding geographers, all attest to the widespread influence, roles and competence of those inspired by the people honored. The list of such contributors to the growth of geography includes, for example, in the United Kingdom, Alan G. Ogilvie (Miller and Watson, 1959), Austin Miller (Whittow and Wood, 1965), and K.C. Edwards (Osborne, Barnes and Doornkamp, 1970); in Australia, Gilbert J. Butland (Douglas, Hobbs and Pigram, 1975), and O.H.K. Spate (Jennings and Linge, 1980); and in the United States, Raymond E. Murphy (Ryan, 1971), and Chauncy D. Harris (Demko and Fuchs, 1984). While not exhausting the list of tributes to outstanding geographers, these volumes illustrate the continuing respect and admiration felt by former students for those who inspired them to seek their own place in the diverse subjects and functions of professional geography.

The present volume in addition to its formal analytical presentations, offers a succession of photographs reflecting different moments in the life of J. Lewis Robinson, at graduate school, in World War II, and individually or with colleagues during his career at the University of British Columbia (1946-1984). The volume concludes with biographical details of Lewis Robinson and a bibliography of his writings arranged by subject and/or area. When requesting these items from Professor Robinson, the editor in the belief that many of the readers of this volume will have been students in one or more of his courses at U.B.C., also sought a brief assessment of the numbers of students taught during Robinson's career. This information not only attests to Lewis Robinson's contribution to the teaching of geography, but also to the important influence which his department has had on the growth of tertiary education in Canada since he established geography at U.B.C. in 1946.

Robinson's biography, bibliography and teaching/lecturing record reflect his outstanding contribution to geography. While this volume of essays records his achievements and demonstrates the esteem in which he is held by former students, it also presents an opportunity to record how Lewis Robinson himself views his contribution to the study of geography in Canada. The editor believes that the biographical record and professional tribute presented here can be greatly enhanced by inclusion of an autobiographical element which adds greater depth to a career spanning many decades and covering far more facets than we otherwise can hope to address. In keeping with the behavioral approach to many elements of human geography which has added considerable richness to our discipline, the editor asked J. Lewis Robinson to reflect on those aspects of his career which summarize or represent the underlying thrust of the professional activities for which he hopes to be remembered. These reflections, summarized below, add considerable depth not only to our understanding of Robinson as a person and a geographer, but also reveal important dimensions of geography's growth from a fledgling subject to a mature discipline in Canada during Robinson's career.

J. Lewis Robinson Reflects on his Career

The following information has been compiled from several letters by Lewis Robinson to this volume's editor in the fall of 1983. During his 38 years at U.B.C., Robinson lectured for 34 years (being on study leave for four years); in addition, he offered lectures during four or five summer sessions. During this period, approximately 2,300 students enrolled in Robinson's fourth-year course on the regional geography of Canada. Some 2,000 students took a major in geography during this period, and most consulted with Lewis Robinson for advice on their programs. Although lacking a complete record of the numbers of students taught in his introductory courses or in courses on human and political geography, Robinson estimates that, in recent years he has taught between 250 and 350 students per year, and in earlier years, between 500 and 600. Accepting the range in these figures, we can estimate that J. Lewis Robinson has taught and advised over 15,000 students during his career at U.B.C., and has offered courses to several hundred more students at other universities. Although these figures include duplications in enrollments because many of us took more than one course from Robinson, they indicate nevertheless his widespread influence on many students.

Lewis Robinson has also affected the geographical awareness of members of the public through his talks on geographical subjects to teachers, planners, businessmen and others during his career. While building a place for geography in the university and in society, Robinson often lectured at least once a week to groups outside the university, averaging between 25 and 30 talks per academic year. Still lecturing at least once a month to public groups, Robinson estimates that, during his career at U.B.C., he has reached somewhere between 15,000 and 20,000 people in this manner.

J. Lewis Robinson in 1943 became the first professional "geographer" to be employed by the Federal Government of Canada. Working under contract for the Northwest Territories Administration, his official title included the word, geographer. One of his duties being to "assemble and coordinate physical, economic and human information about the Northwest

Territories," Robinson spent four summer field seasons in the Territories, wrote one booklet, and prepared a series of articles on the area for the *Canadian Geographical Journal*, from which reprints were distributed by the Canadian government as official information.

In 1946, President Norman Mackenzie of U.B.C. directly invited Lewis Robinson to move to Vancouver and establish a geography program within the university's Department of Geology and Geography. Receiving his Ph.D. degree from Clark University at the time of his appointment to U.B.C., Robinson was appointed with the rank of Associate Professor; in 1953 he became the first chairman of a Geography Division within the joint department, and became the first head of the Geography Department when it was established as a separate administrative unit in 1959.

Lewis Robinson became one of the pioneers in British Columbia in the promotion of regional planning. As part of a team of four persons, during the course of several years he toured the province for the Community Planning Association of Canada speaking about principles and advantages of planning, and promoting the value and purpose of geography as well. Several of his department's graduates were among the first to be employed as professional planners in British Columbia.

Following the example of his mentor and role model at the University of Western Ontario, Professor Edward Pleva who encouraged his leaving history and entering geography, Robinson participated in curriculum committees in the 1950s and 1960s to promote geography and social studies programs in British Columbia schools, and spoke annually at numerous teachers' conventions throughout the province. At that time, nearly half of Robinson's upper year geography students subsequently became teachers in British Columbia schools.

Following his retirement as Head of the Geography Department in 1968, Lewis Robinson undertook to focus on the department's undergraduate programs as Undergraduate Advisor, a position he held for fifteen years. As teacher and guide of students, Robinson looks "back on the 38 years with pride and satisfaction"; he adds, wistfully, that "only they, the students, know what impact I had on their lives."

Lewis Robinson remembers his first twenty years at U.B.C., when the number of graduate students was much smaller than it later became, as a time when he and his wife, Jo, met incoming graduate students at the airport or train station, grew to know them through dinners and parties in their home, and maintained close contact when they left U.B.C. to become professional geographers. Still important among the Robinsons' network of personal friends, the relationships with geographers begun in U.B.C.'s graduate program signify to Lewis Robinson an important "part of the 'family atmosphere' which made the U.B.C. Geography Department a happy place."

Lewis and Jo Robinson have also had a great impact on the lives of the deaf in British Columbia. He writes that, "as a result of our oldest daughter being born deaf we dedicated our non-academic lives to helping children and adults with their 'unseen' handicap. We organized the first Parents' Association at the provincial school for the deaf and Jo and I each served as presidents of it, twice." When their daughter became a champion swimmer, Lewis Robinson helped to organize the first Trans-Canada Federation of Silent Sports and served for several years on its board of directors. This same daughter is now president of what has come to be known as Deaf Sports in Canada. The Robinsons helped establish the Western Institute for the Deaf which helps the adult deaf in Vancouver, and each served as presidents of that organization, and in many capacities as members of its executive. Jo Robinson is still active as a volunteer helper in the institute.

Perspective

Building on the biographical, bibliographical and introductory components of this volume, the contributors hope that their essays in honor of J. Lewis Robinson will be accepted as partial acknowledgment of his professional and academic record in the field of Canadian geography during the past four decades. Now dispersed throughout Canada, the authors represented in this volume continue to feel great empathy with Professor J. Lewis Robinson and strive to emulate the role-model which he has so faithfully represented.

REFERENCES

DEMKO, G. and FUCHS, R. (1984; forthcoming), *Festschrift for Chauncy D. Harris* (Chicago: University of Chicago, Department of Geography Research Papers).

DOUGLAS, I., HOBBS, J.E. and PIGRAM, J.J. (1975), *Geographical Essays in Honour of Gilbert J. Butland* (Armidale: University of New England, Department of Geography).

JENNINGS, J.N. and LINGE, G.J.R. (1980), *Of Time and Place. Essays in Honour of O.H.K. Spate* (Canberra: Australian National University Press).

MILLER, R. and WATSON, J.W. (1959), *Geographical Essays in Memory of Alan G. Ogilvie* (London: Thomas Nelson and Sons Ltd.).

OSBORNE, R.H., BARNES, F.A. and DOORNKAMP, J.C. (1970), *Geographical Essays in Honour of K.C. Edwards* (Nottingham: University of Nottingham, Department of Geography).

RYAN, C.J. (Guest Editor) (1971), *Festschrift for Raymond E. Murphy, Economic Geography,* 47(2).

WHITTOW, J.B. and WOOD, P.D. (1965), *Essays in Geography for Austin Miller* (Reading: University of Reading).

Lecturing in the Geography of Canada Course, U.B.C., 1981

1

THE FIORDLANDS OF NORTHEASTERN CANADA

Robert Gilbert

Approximately four times more fiords are found in the Arctic than on the west coast of North America, but arctic fiords on average are significantly smaller. Furthermore, while the drainage basins of the largest arctic fiords are larger than those on the west coast, their lower runoff provides a significantly smaller flow of fresh water per unit area of fiord surface. Without the strong stratification produced in west-coast fiords by such a freshwater cap, the waters of arctic fiords circulate freely. This circulation has important implications for sedimentation, biological processes, and the nearshore impact of human activities.

FIORDS

The east coast of Canada from central Labrador to northern Ellesmere Island is dominated by fiords. But, while the fiords of western Canada have been extensively studied for many years (for example, Pickard and Stanton, 1980), our understanding of the oceanography, sedimentology, and biology of east coast fiords until recently has been very much poorer. Even the bathymetry of major fiords is unknown. Lind and Andrews (in press),for example are able to discuss bathymetric data from only 19% of the Baffin Island fiords, almost none of which have been completely sounded.

The major studies in fiords of eastern Canada prior to 1970 had been conducted by the Department of National Defence in the Queen Elizebeth Islands (Hattersley-Smith, 1965; 1967; Ford and Hattersley-Smith, 1965), the Geographical Branch of the Department of Mines and Technical Surveys (Loken, 1966), and the *Blue Dolphin* Expeditions to Labrador (Nutt, 1951; 1963). During the past decade, however many studies have been carried out in association with assessment of the environmental impact of potential oil and gas exploitation in the eastern Arctic and offshore Labrador. The results of the Eastern Arctic Marine Environmental Studies Program (some of which were published in a special edition of Arctic, 35 (1), 1982) are examples of this work.

Representing a vital component of the interface between the land and the sea, the fiords have been the focus of increasing attention (Lake and Walker, 1976), which culminated with an international, multidisciplinary research effort in 1982 and 1983 under the auspices of the Bedford Institute of Oceanography (Syvitski and Blakeney, 1983).

This paper briefly reviews from a geographical perspective our knowledge of eastern arctic and subarctic fiords, and assesses the needs for further study.

A COMPARISON OF THE SIZE AND HYDROLOGY OF EAST- AND WEST-COAST FIORDS

The data summarized in Table 1 facilitate an assessment of the number and size of east-coast and west-coast fiords. Because fiords, bays and estuaries comprise a continuum of form, distinction among them is arbitrary. Fiords defined here as embayments longer than 8 km whose length is greater than three times their width, correspond to those defined on the west coast by Pickard and Stanton (1980). If shorter embayments were considered, then the number of east-coast fiords would be much larger. Lind and Andrews (in press), for example report 227 fiords on Baffin Island between 5 to 227 km long and document 141 from 8 to 273 km (including Admiralty Inlet). Fiords are defined in addition as having terrain rising above 328 m within 2 km of the water. This distinction is made in the absence of bathymetric data to distinguish fiords from shallow embayments and estuaries. Fiords with tributaries are considered single fiords if one distinct mouth at the seaward end can be defined, even though their tributaries may themselves be larger than other fiords evaluated separately (as for example, those of Greely Fiord).

Approximately four times more fiords exist in eastern North America than on its west coast (Table 1) and they occur over a wide range in latitude, 52° to 83°N (Figure 1). On average, however, only the fiords of the northern region including north Baffin, and Ellesmere are comparable in length to those of the west coast. In part due to their location

TABLE 1

COMPARISON OF LENGTHS OF EAST- AND WEST-COAST FIORDS

Region	Mean Length (km)	Std. dev. (km)	Max. length (km)	No. longer than 100 km	50 km	20 km	Total number
West Coast[1]	53.9	56.6	380	13	27	70	85
Alaska[1]	115.9	94.5	380	8	11	16	16
British Columbia[1]	47.5	32.4	130	5	14	40	44
Vancouver Island[1]	25.4	16.0	70	-	2	14	25
Northeastern Coast	32.1	29.0	273.	9	54	170	306
Labrador-Ungava	22.4	14.4	56.	-	4	19	44
South Baffin[2]	21.2	14.9	70.	-	2	12	39
Cumberland Pen.[3]	25.1	15.7	83.2	-	2	22	45
North Baffin[4]	46.9	40.4	273.	3	23	42	57
Devon Island	22.6	12.1	45.	-	-	11	22
Ellesmere east[5]	36.1	24.4	101.	2	6	20	30
Ellesmere west	49.0	45.2	191.	4	10	26	35
Axel Heiberg Island	26.0	16.1	56.5	-	4	12	23
Other[6]	29.3	19.5	65.5	-	3	6	11

[1]Data from Pickard and Stanton (1980).

[2]From southeastern Frobisher Bay to the head of Cumberland Sound.

[3]From the head of Cumberland Sound to Broughton Island.

[4]North of Broughton Island.

[5]Including the south coast, north to Cape Aldrich.

[6]Melville Island, Somerset Island, and Victoria Island.

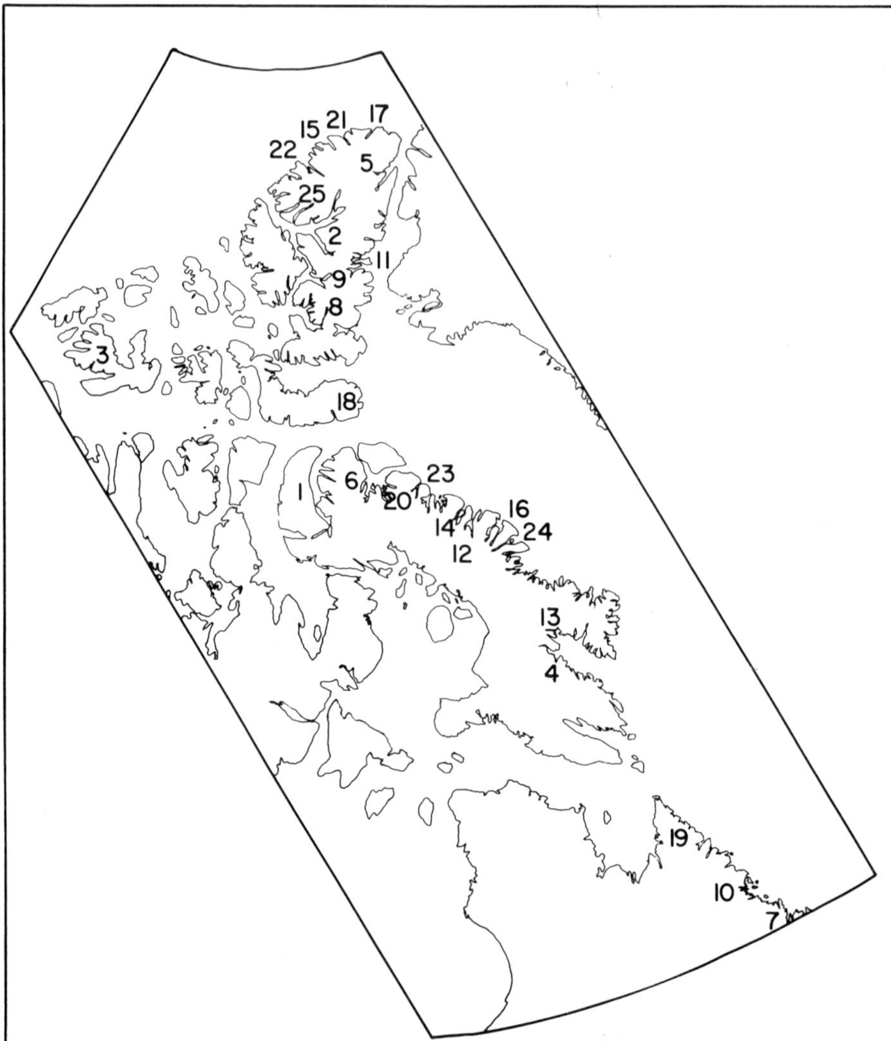

Figure 1. The Fiordlands of Northeastern Canada
(Numbers correspond to entries in Table 2)

on the arctic islands and on the relatively narrow raised rim of the Canadian Shield, only 9 fiords in eastern Canada are longer than 100 km compared with 13 on the west coast.

The most important factor in the oceanography of fiords, the frequency and extent of circulation of their waters and of their exchange with the open sea beyond, significantly controls the sedimentary processes within the fiord and the existence of biota. A major factor controlling circulation is the amount and timing of freshwater flow into the fiord. Generally, large inflows of fresh water create a cap on the surface of the fiord inhibiting mixing across the boundary (pycnocline) with salt water below. The cap is particulary effective if the layer of freshwater extends to the sill and blocks exchange with the ocean beyond.

This characteristic differs in general between west-coast fiords with high runoff and east-coast fiords with lower total runoff and a period of very low or zero inflow. All east-coast fiords which have been studied to data are known to circulate freely, the only exceptions being bays almost cut off from the sea by nearly emergent sills.

Because runoff data are almost completely absent for fiord regions on the east coast, a comparison has been made in Table 2 among the fiords with the largest drainage basins. Comparison of the largest fiords of both coasts reveals that those of the east coast have significantly larger basins and areas of water surface. Because the runoff is much less, however, especially to the larger northern fiords, the depth of water placed on fiord surfaces each year is much smaller than on the west coast. Except for several small fiords receiving inflow from very large basins (Irvine, Conybeare, Clearwater and Croker) and some of the fiords of Labrador (Ugjoktok and Nain-Tikkoatokak) where inflow is greater, the depth of water added is not more than several meters per year. In west-coast fiords, values are one to two orders of magnitude greater. Furthermore except in the most southerly regions or where a small amount of groundwater flow may be occurring through sandurs--described in Cambridge Fiord, northern Baffin Island by Sadler and Serson (1980) and Hay (1983)--no inflow occurs on much of the east coast

TABLE 2

RUNOFF FROM THE FIORDS WITH LARGEST DRAINAGE BASINS

ON THE EAST- AND WEST-COASTS OF CANADA

Fiord	Area (km^2) Drainage basin	Area (km^2) Water surface	Runoff[1] (m/a)	Runoff per unit area of fiord (m)
West Coast				
Dean Channel	10995	142	2.8	220
Bute Inlet	9476	243	1.5	60
Knight Inlet	8425	251	1.9	66
Gardner Canal	6793	201	2.8	97
Howe Sound	4773	285	1.8	32
Rivers Inlet	4700	150	2.6	84
Toba Inlet	2764	89	1.8	57
Alberni Inlet	2323	73	2.7	88
Jervis Inlet	1877	194	1.9	20
S. Bentinck Arm	1836	92	2.5	52
Arctic and Subarctic[2]				
1. Admiralty Inlet	44100	9075	0.10	0.59
2. Greeley Fiord	32300	5844	0.10	0.65
3. Minto Inlet	22400	1517	0.09	1.4
4. Irvine Inlet	14000	60	0.30	70.
5. Conybeare Fiord	12800	221	0.15	8.8
6. Milne Inlet	11000	1231	0.11	1.1
7. Ugjoktok Bay	10800	85	0.65	83.
8. Bauman-Vendom Fiord	10300	2009	0.10	0.61
9. Bay Fiord	7510	1036	0.13	1.1
10. Nain-Tikkoatokak B.	7470	203	0.63	24.
11. Makinson Fiord	7360	946	0.11	0.97
12. Sam Ford Fiord	6460	815	0.15	1.3
13. Clearwater Fiord	5930	156	0.30	12.
14. Gibbs Fiord	5540	308	0.13	2.5
15. Yelverton Inlet	4960	283	0.12	2.2
16. Clyde Fiord	4840	537	0.20	2.0
17. Clements Markham F.	4720	306	0.16	2.6
18. Croker Inlet	4690	356	0.15	2.1
19. Hebron Fiord	4680	211	0.55	13.
20. Tay Sound	4580	248	0.13	2.5
21. M'Clintock Inlet	4510	356	0.12	1.6
22. Phillips Inlet	4280	411	0.12	1.4
23. Coutts-North Arm F.	3610	110	0.16	5.4
24. McBeth Fiord	3580	402	0.25	2.5
25. Otto Fiord	3250	778	0.13	0.67

[1]Estimated from data in the Hydrological Atlas of Canada (1978).

[2]See Figure 1 for the location of these fiords.

for periods of up to 8 months during winter. Circulation of arctic fiords consequently is much more vigorous than that of fiords on the west coast.

The melting each spring of a two-meter thick layer of sea ice with an average salinity of 5% contributes a layer of nearly fresh water some 1.8 m thick, comparable in thickness to that contributed by flow into a number of fiords and significantly greater than than into others. This brackish layer produced before breakup and the presence of sea ice throughout the winter, prevent mixing by wind and serve as stabilizing influences. They do not, however, deter deep mixing by tidal forces in the weakly stratified water.

The circulation resulting in part from these hydrologic conditions has a number of significant consequences. Oxygen and nutrient-rich waters are brought almost continuously to the floor of the fiord and an extensive benthos develops. Molluscs, marine worms, brittle stars, anthropods, bryzoans, anemones, sea urchins and other organisms are common on the floors of fiords on Baffin Island (Gilbert, 1978, 1982; Farrow, 1983). The role of these organisms in the food chain including the sea mammals of the region is largely unknown. From the point of view of the sedimentologist, the effect of these abundant and active organisms is to bioturbate the sediments severely. In some locations the otherwise valuable record of the Quaternary environment of the fiord is destroyed.

Calving glaciers and icebergs in the arctic fiords also induce circulation (Greisman, 1979). Lower density fresh water entering at the base and along the faces of these ice walls entrains the sea water as it rises to the surface. Ambient water from the upper layers replaces the sea water at depth. Nutrients are provided from the fine glacial sediments, and ice proximal, benthic biota are especially abundant in spite of the rather severe physical conditions at the calving faces. Sea birds and marine mammals are often observed in these areas in greater abundance than elsewhere in the fiord and attest to the richness of the benthos.

The circulation of the deep waters of eastern fiords means that disposal of wastes in fiord waters such as that occurring in western Canada (Hay et al., in press), presents significant problems. Materials such as toxic mine wastes cannot be deposited on the fiord floor without the significant risk of their entering the food chain either by assimilation on the fiord floor, or by resuspension, dissolution and transportation to the upper layers and possibly to the ocean beyond. Questions related to such processes have been raised by the disposal of tailings at the Nanisivik Mine on Strathcona Sound in northern Baffin Island (Hatfield and Williams, 1976). In Greenland various scenarios for the containment of tailings from the Black Angel Mine in Aqfardlikavsa Fiord have been proposed (Nielsen and Hansen, 1980), and some understanding of arctic fiord oceanography has been reached there through extensive studies of circulation (Asmund, 1980; Lewis and Perkin, 1982).

CONCLUSIONS:
FOCI OF ONGOING AND FUTURE STUDIES OF ARCTIC FIORDS

Our understanding of arctic oceanography and marine sedimentology has developed almost entirely since 1945. However, it has been in the last decade that large, multidisciplinary research projects have produced the most significant results. Much of the work has been motivated by the need to acquire basic information for environmental assessment of the impact of resource exploitation, especially in the waters among the Canadian arctic islands.

As the information in Tables 1 and 2 indicates, fiords are the most extensive single feature of the Canadian coast of the Arctic Ocean. They control the distribution of a significant portion of the sediment presently being eroded from arctic lands, and their waters and biota interact with those of the open ocean. Yet until recently, our understanding of the sedimentary and biologic processes of arctic fiords was fragmentary.

The Bedford Institute of Oceanography in 1982 began a major two year program in ten fiords of Baffin Island to assess the nature of sedimentation. Although several years will elapse before the large volume of data from the study is analysed completely, many important areas of research have yet to be investigated.

In the field of arctic fiord sedimentology, perhaps most important for research is the study of the processes of sediment erosion and deposition by mass wasting and gravitational flow. We know from examination of side scan sonar records of fiord walls that large quantities of sediment are redistributed by both rapid and slow slope failure. Models of these processes have been proposed (Gilbert, in press) but further study is required to assess their mechanics, their frequency and magnitude, and their causes. Although large trenches observed in the floors of some fiords possibly are eroded by turbidity currents generated by massive slope failures, no observations have been made as the process occurs. Studies are just beginning of the deposits from these events and their effect on the benthos. Little is known about the possible impact on the environment expected from the use of fiords and nearby oceans for exploiting and shipping resources. Important questions remain to be answered about the oceanography of fiord waters, their circulation, and their communication with the open ocean. The distribution of pollutants, and the interaction of biota throughout the nearshore of arctic waters, are of special concern.

Studies of the origin of arctic fiords comprise an interesting specific field of inquiry which relates to many broad areas of investigation including major tectonic events such as the drifting of Greenland from the North American land mass, the role of pre-Pleistocene fluvial processes and landscape evolution, and the importance of glacial erosion in the enlargement and modification of fiord form. Many of the studies reviewed in this evaluation together represent a useful first step in expanding our knowledge of the origins and characteristics of fiords in northeastern Canada.

REFERENCES

ASMUND, G. (1980), Water movements traced by metals dissolved from mine tailings deposited in a fjord in north west Greenland, in H.J. Freeland, D.M. Farmer and C.D. Levings (editors), *Fjord Oceanography* (New York: Plenum Press): 347-353.

FARROW, G.E. (1983), Bottom fauna and Bioturbation, in J.P.M. Syvitski and C.P. Blakeney (compilers), *Sedimentology of Arctic Fjords Experiment: HU 82-031 Data Report*, Volume 1 (Ottawa: Data Report, Hydrography and Ocean Sciences No. 12): 9-1 to 9-25.

FORD, W.L. and HATTERSLEY-SMITH, G. (1965), On the oceanography of the Nansen Sound fiord system, *Arctic*, 18: 158-171.

GILBERT, R. (1978), Observations on oceanography and sedimentation at Pangnirtung Fiord, Baffin Island, *Maritime Sediments*, 14: 1-9.

GILBERT, R. (1982), Contemporary sedimentary environments on Baffin Island, N.W.T., Canada: Glaciomarine processes in fiords of eastern Cumberland Peninsula, *Arctic and Alpine Research*, 14: 1-12.

GILBERT, R. (in press), Sedimentary processes of Canadian arctic fiords, *Sedimentary Geology*.

GREISMAN, P. (1979), On upwelling driven by the melt of ice shelves and tidewater glaciers, *Deep Sea Research*, 26A: 1051-1065.

HATFIELD, C.T. and WILLIAMS, G.L. (1976), *A summary of Possible Environmental Effects of Disposing Mine Tailings into Strathcona Sound, Baffin Island* (Ottawa: Indian and Northern Affairs, ALUR 75-76-38): 76 pp.

HATTERSLEY-SMITH, G. (1965), Canadian operation "Hazen Tanquary," 1964, *Polar Record*, 12(80): 592-595.

HATTERSLEY-SMITH, G. (1967), Canadian operation "Hazen Tanquary," 1967, *Polar Record*, 13(86): 607-610.

HAY, A.E. (1983), Cambridge Fjord polynya experiment: The buoyant jet, in J.P.M. Syvitski and C.P. Blakeney (compilers), *Sedimentology of Arctic Fjords Experiment: HU 82-031 Data Report*, Volume 1 (Ottawa: Data Report, Hydrography and Ocean Sciences No. 12): 7-1 to 7-4.

HAY, A.E., MURRAY, J.W. and BURLING, R.W. (in press), Submarine channels in Rupert Inlet, B.C.: I. Morphology, *Sedimentary Geology*.

LAKE, R.A. and WALKER, E.R. (1976), A Canadian arctic fjord with some comparisons to fjords of the western Americas, *Journal of the Fisheries Research Board of Canada*, 33: 2272-2285.

LEWIS, E.L. and PERKIN, R.G. (1982), Seasonal mixing processes in an arctic fjord system, *Journal of Physical Oceanography*, 12: 74-83.

LIND, E.K. and ANDREWS, J.T. (in press), The fiords of Baffin Island: Description and classification, in J.T. Andrews (editor), *Quaternary Studies on Baffin Island, Baffin Bay and West Greenland* (London: Allen and Unwin).

LOKEN, O.H. (1966), Science on Baffin Island: Field research in physical geography, *Canadian Geographical Journal*, 72: 38-47.

NEILSEN, T.K. and HANSEN, N.-E.O. (1980), Mixing and exchange processes in a small Greenland sill fjord, in H.J. Freeland, D.M. Farmer and C.D. Levings (editors), *Fjord Oceanography* (New York: Plenum Press): 219-225.

NUTT, D.C. (1951), The *Blue Dolphin* Labrador expeditions, 1949 and 1950, *Arctic*, 4: 1-11.

PICKARD, G.L. and STANTON, B.R. (1980), Pacific fjords - a review of their water characteristics, in H.J. Freeland, D.M. Farmer and C.D. Levings (editors), *Fjord Oceanography* (New York: Plenum Press): 1-51.

SADLER, H.E. and SERSON, H.V. (1980), An unusual polynya in an arctic fjord, in H.J. Freeland, D.M. Farmer and C.D. Levings (editors), *Fjord Oceanography* (New York: Plenum Press): 299-304.

SYVITSKI, J.P.M. and BLAKENEY, C.P. (compilers) (1983), *Sedimentology of Arctic Fjords Experiment: HU 82-031 Data Report*, Volume 1 (Ottawa: Data Report, Hydrography and Ocean Sciences No. 12): 935 pp.

2

WATER BALANCE STUDIES IN THE CANADIAN ARCTIC ARCHIPELAGO

Ming-ko Woo

Although Arctic hydrology has advanced significantly in the past decade, inaccurate and insufficient data previously prevented computation of annual water balance for most parts of the Canadian Arctic Islands. Based on a six-year study, the magnitude of various water balance components was found to vary greatly from year to year. A comparison of five similar studies carried out in the non-glacierized areas of the Arctic archipelago shows that regional differences in water balance reflect a spatial change in precipitation; runoff usually consumes the bulk of annual precipitation except where wetland and open water conditions prevail. Further work on the regional water balance of the Arctic Islands will have practical applications and will improve our knowledge on the hydrology of Canada's last terrestrial frontier.

WATER BALANCE

The development of the Canadian Arctic Islands includes megaprojects such as the opening of mines, seaports and gas pipelines which involve problems of water supply or risks of floods, erosion and sedimentation. An assessment of the various hydrological quantities is fundamental to rational planning of water use and the prevention of water-related hazards. One unifying characterization of the hydrological processes in a geographical region is the equation for water balances which evaluates for a given period the amount of water gained, lost and stored in a drainage basin. The equation is:

$$ds/dt = M + R - E - Q$$

where M is snowmelt, R is rainfall, E is evaporation, Q is runoff, and ds/dt is the rate of change of storage. This chapter assesses the problems of evaluating the water balance for the Arctic Islands, and attempts to regionalize the pattern of water balance by summarizing the results of a limited number of studies.

Inaccurate and insufficient data prior to 1970 prevented annual water balances from being calculated. Hare and Hay (1971) thus excluded the Arctic Islands from their analysis of Canada's annual water balance, and Lvovitch (1973) similarly omitted the Canadian Arctic archipelago from his study of global water balances. The difficulty of acquiring basic hydrological information such as streamflow and evaporation is partly due to the remoteness of the region. The processes operating in an environment of intense and prolonged coldness where permafrost, snow and ice exert prominent influences upon the hydrological cycle are insufficiently understood. The lack of process studies hinders interpretation of field observations, making it difficult to extend either in space or in time the results of any one study.

The amount of work has increased since 1970 on various elements of the Arctic hydrological system, including precipitation (Cogley and McCann, 1976; Woo and Marsh, 1978), snowmelt processes (Heron and Woo, 1978), evaporation (Addison, 1977; Ohmura, 1982a), slope runoff (Lewkowicz and French 1982; Woo and Steer, 1982), and streamflow (McCann and Cogley, 1972; Woo and Sauriol, 1981). Advances made by such research, through significant, did not overcome the difficulties of computing water balances because the short-term nature of the studies did not enable representative mean values of the water balance components to be obtained (Ohmura, 1982b; Woo et al., 1981). Furthermore, site specific investigations often focus upon single processes and thus prevent a regional water balance from being established.

Discussions in this chapter are limited to the non-glacierized coastal basins of the Arctic Islands where most of the data have been collected.

ARCTIC ISLANDS ENVIRONMENT

The Canadian Arctic archipelago extends from the south of Baffin Island (61°N) to the northern tip of Ellesmere Island (83°N), and from longitude 60°W to 126°W (Figure 1). The eastern flank is extremely rugged, with mountains often rising abruptly from sea level to over 2,000 meters. Many higher areas are capped by glaciers whose tongues along

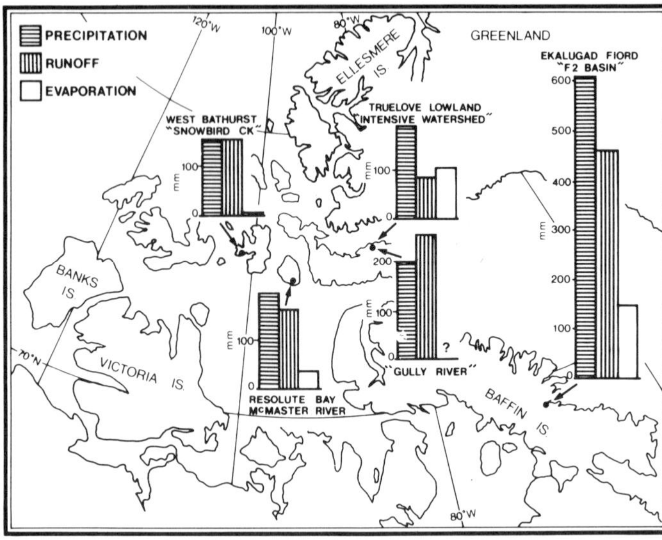

Figure 1. Canadian Arctic Archipelago and Location of Five Water Balance Study Basins. The Magnitudes of Precipitation, Runoff, and Evaporation for these Basins are also shown.

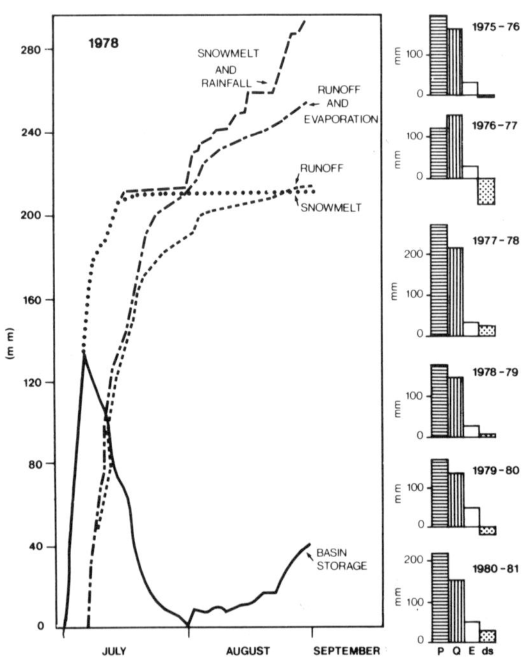

Figure 2. Daily Water Balance for McMaster River Basin during the Summer of 1981; and Annual Water Balances for 1975-81. (P = precipitation; Q = runoff; E = evaporation; ds = change in storage)

major valleys sometimes reach sea level. The central section, the Arctic lowland region (Bostock, 1970), is occupied by dissected plateaux. The western fringe is the Arctic coastal plain, a lowlying area that slopes gently northwestward towards the continental shelf. Climatically, the region has temperatures below freezing lasting 8 to 10 months each year which maintain continuous permafrost, often at a depth of less than 1 m. Snow cover lasts for 9 months, and lake ice can grow to an annual maximum thickness exceeding 2.5 m. During the short summer, rainfall is common in the lowlying areas. Weather stations report an annual precipitation exceeding 500 mm in the eastern Arctic but decreasing sharply towards the west.

A combination of low temperature and low precipitation gives rise to polar desert conditions. Surface vegetation is sparse in the north but some tundra is concentrated in sheltered and wetter areas. Vegetation becomes more continuous southwards, though the entire area is north of the tree line.

The preponderant influence of snow on Arctic hydrology has been much emphasized (Cook, 1967; Church, 1974; Woo, 1983). Windswept winter conditions lead to an uneven distribution of snow, with shallower snowpacks on exposed, flat terrains, and deeper snowpacks in depressions such as gullies and valleys (Woo *et al*., 1983). Snowmelt begins in late May in the south but is delayed until mid-June in the north. The bulk of the snow cover in most years is depleted in two or three weeks. A rapid release of meltwater invariably produces large peaks of runoff. Runoff declines after the melt season and only major rainstorms produce significant high flows. By mid-September or October, streamflow ceases and winter begins.

PROBLEMS OF WATER BALANCE ESTIMATES

Although the general features of the hydrological regime are well documented, a long-term and extensive data collection network is lacking. This, together with various problems of measurement has hindered the estimation of water balance of Arctic drainage basins.

Precipitation measurements in the Canadian Arctic have suffered inaccuracies because of poor catch efficiency of gauge under windy conditions, and because of the frequent trace events which cannot be recorded by conventional gauges (Jackson, 1960; Woo and Steer, 1979). Snowfall is particularly prone to large errors of measurement. Woo et al. (1983), for example, found that the total snowfall recorded at weather stations represented only about 40 to 60 per cent of the winter snow accumulation in nearby drainage basins. Orographic effects can also be pronounced in rugged terrain (Ohmura, 1982b) but they are difficult to determine quantitatively. These factors usually result in considerable underestimation of basin precipitation if the weather station data are used to calculate the basin water balance. One way to overcome this problem is to conduct an intensive basin-wide snow survey at the end of winter (Wedel et al., 1977; Woo and Marsh, 1978). Because the snowpack does not undergo any melting throughout the long polar winters, the survey should provide a good measure of total snow accumulation. By traversing various terrains found in the basin, the survey should produce a satisfactory estimate of the mean snow conditon.

Unlike precipitation which should be observed at many points in the basin and then averaged, runoff can be obtained at a single point by determining the discharge at the basin outlet. Conventionally, stream level (or stage) is continuously monitored and the stage record is converted into discharge using empirically derived stage-discharge relationships. Being empirical, such relationships change when the channels become unstable, a condition frequently observed during breakup. At that time because meltwater flows along valleys may be deeply infilled with snow (and aufeis in more southerly locations), channels tend to shift in the snow and ice (Pissart, 1967; Woo and Sauriol, 1980). The morphology of the channel alters continually, making it difficult to establish reliable stage-discharge rating curves. To overcome this problem, streamflow should be directly measured during the snowmelt period, although because this can be a hazardous operation with fast flow through unstable channels lined with snow and ice, errors of measurement may be considerable.

Evapóration is usually determined for a single site and the computed values are assumed to be applicable to the entire basin. This assumption is not valid for the snowmelt period, because when the basin snow cover breaks up into patches, only the snow-free zones undergo evaporation and the snow itself may or may not experience sublimation loss. As spring progresses, the basin may dry out differentially and evaporation rates will be affected by the spatial variation of moisture available at the ground surface. To assume therefore a potential evaporation rate for drainage basins which are relatively dry in summer is erroneous (Bovis and Barry (1974) pointed out this limitation in their estimation of regional evaporation for the Arctic). Before extending the point estimates of evaporation obtained by lysimeter measurements or by microclimatic methods (e.g., Addison, 1977), basin conditions must be assertained which are similiar to those of the site where evaporation has been measured.

Sublimation may be substantial in certain parts of the Arctic Islands. Ryden (1977) computed that a quarter of the winter snowpack at Truelove lowland, northern Devon Island, was lost to sublimation. For most other areas, the amount of water loss through this process is small, though the equivalent energy loss may be considerable (because of the large latent heat involved in sublimation).

Change in basin storage is difficult to measure directly. For simplicity, many workers have either assumed that the annual change is negligible (Marsh and Woo, 1979), or have computed storage change as a residual term of the water balance (e.g., Holecek and Vosahlo, 1975). Because annual storage change may actually not approach zero, the storage component if left as a residual term inevitably incorporates the inaccuracies arising from determination of other components of the water balance. Field measurement to date has not determined the year-to-year change in basin storage.

AN EXAMPLE OF BASIN WATER BALANCE

Most water balance studies of Arctic basins have been carried out for one to three years. The only exception is a six-year study made for McMaster basin located near Resolute, Cornwallis Island (74°45'N, 94°50'W), used here to illustrate the water balance of a small Arctic drainage basin.

McMaster basin (area 33 km^2) is centrally located in the Arctic archipelago and has elevations ranging from 80 to 200 m above sea level. The basin has a rolling topography, dominant surface materials of gravel and loam, and a scattered tundra vegetation cover. Continuous permafrost, found at 0.2 to 0.7 m below the surface, is maintained by prolonged sub-freezing air temperatures. The snow cover lasts for about nine months each year, and such hydrological activities as runoff and evaporation are limited to the period between late June and early September. Details regarding field measurement and computation methods for water balance are given in Woo,Marsh, and Steer (1983) and need only be summarized here.

The basin was divided into several terrain units; snow surveys were carried out at the end of each winter to determine the toal accumulation and distribution of snow in various types of terrain. Snowmelt at each terrain unit was calculated daily in spring, using an energy-balance approach (Heron and Woo, 1978), and summed to yield total melt for the basin. Because the snow distribution was uneven at the end of winter, the snow cover quickly broke into patches as melting began. When a certain terrain unit became free of snow, the magnitude of resulting evaporation depended upon the available amount of energy and surface soil moisture (Marsh et al., 1981). Rainfall was measured by a network of gauges and streamflow was obtained at the outlet of the basin. Water balance computations were made on a daily basis, and a "water year" was considered to span the time between the first significant snowfall in September of the first year (when the snow would stay for the winter), and the arrival of snowfall a year later.

The cumulative daily snowmelt, rainfall, evaporation, runoff and change in storage (computed as a residual) were recorded for the 1978 field season (Figure 2). Snowmelt began late that year (July 1) but three-quarters of total winter snow accumulation was depleted within one week. Evaporation was negligible until runoff began on July 6. With the release of a large amount of meltwater, but a comparatively low amount of water loss, a large volume of water was stored in the basin. Field observation showed that the bulk of this storage was retained first in the deep valley snowpacks which became saturated with water melting on the slopes. As the ground began to thaw, an increasing amount of water was stored in the soil. Peak runoff occurred once an integrated channel flow network was established in the snow-filled valleys (Woo and Sauriol, 1980) to drain most of the basin's stored water. Evaporation, though low in magnitude, increased gradually. Storage was much reduced by late July and dry conditions became prevalent. Rainfall in August occasionally raised the streamflow but a sizable portion of the precipitation went into replenishing the basin's storage. Streamflow and evaporation declined by September, as freeze-up began. An excess of storage at the end of summer extended into the subsequent spring, showing that the net change in storage at the end of the year is not negligible.

Various components of the annual water balance for 1975-1982, including precipitation, runoff, evaporation and net storage change have been calculated (Figure 2). Runoff in all years, accounted for a large part of annual precipitation. Year-to-year fluctuations in precipitation and runoff were noticeable but variation in the magnitude of evaporation was the most conservative. The computed annual change in storage also varied and showed an excess in some years, and a deficit in others. Over six years, annual precipitation averaged 190 mm, runoff 158 mm, and evaporation 38 mm. The computed net storage deficit for the entire study period was 34 mm.

REGIONAL WATER BALANCE

Results of the McMaster basin study may be compared with other findings to derive a pattern of regional water balance for the Arctic Islands. Although earlier attempts made use of available precipitation and runoff data, and computed evaporation from empirical formulae (e.g., Walker and Lake, 1975; Bovis and Barry, 1974), the accuracy of their data cannot be verified. To ensure greater confidence in the following discussion, only those basin studies are included which directly measured at least two of the water balance components. The five basins of the unglacierized areas of the Arctic Islands, which qualify are: "F2" watershed of eastern Baffin Island; "Intensive" watershed and "Gully River" basin of Truelove Inlet, northern Devon Island; "Snowbird Creek" basin of western Bathurst Island; and McMaster basin of southern Cornwallis Island (Table 1 and Figure 1).

The long-term change in basin storage was assumed or calculated to be small, except for "Gully River" whose water balance computation indicated a large water deficit for 1978. Holecek and Vosahlo (1975) argued that water released from ground ice storage could balance the deficit, but the accuracy of their original data is suspect. Ryden (1977) suggested that precipitation was underestimated for this basin because the snow survey left out an unknown quantity of snow stored in the deep ravines.

The six-year study at McMaster basin demonstrated that the annual water balance can vary greatly from year to year (Figure 2); as indicated by Table 1, however, the field seasons of all the other studies comprised only one to three years.

Anomalously high or low precipitation for any study year can render the data atypical of long-term average conditions. Despite such limitations, however, precipitation for "Snowbird" basin, Truelove Lowland and McMaster basin was comparable, approximately 200 mm per year. This low value confirms that many parts of the Arctic Islands are polar deserts (Bovis and Barry, 1974). The much higher value for "F2" watershed at Ekalugad Fiord is readily attributable to the heavy (and often orographic) precipitation of the Eastern Arctic (Maxwell, 1980).

Runoff for all basins except "Intensive" watershed accounts for most of the total precipitation. A high runoff ratio (Q/P exceeding 0.75) is characteristic of the Arctic region above the tree-line. This value exceeds the runoff ratios of many other environments: Sellers (1965) reported a ratio of 0.3 - 0.7 for forests, 0.1 - 0.3 for steppes, and 0.03 - 0.01 for semi-deserts. The large Arctic runoff ratio is mainly due to permafrost near the ground surface, an impermeable layer limiting the basin's storage capacity and encouraging meltwater or heavy rainfall to drain rapidly.

TABLE 1

SUMMARY OF WATER BALANCE FOR FIVE DRAINAGE BASINS IN THE CANADIAN ARCTIC ARCHIPELAGO

Drainage Basin	Location	Area (km^2)	Surface cover	Period of study	Precipitation (mm)	Evaporation (mm)	Runoff (mm)	Reference
"F2" watershed, Ekalugad Fiord	69°N 69°W	2.5	Mainly vegetated bedrock and thin drift	1976	616	154	462	Church (1974)
"Intensive" watershed	76°N	0.12	63% vegetated and bare soil, 12% water, 25% bedrock	1972-74	185	101	84	Ryden (1977)
"Gully River," Truelove Inlet Devon Island	76°N 85°W	22.4	Barren polar desert	1972-73	198	?	250	Holecek and Vosahlo (1975)
"Snowbird Creek," Western Bathurst Island	76°N 101°W	80	Mainly barren polar desert and bedrock, 10% vegetated mainly in lowland	1976	158	?	156	Wedel *et al.* (1977)
McMaster basin	75°N 95°W	33	Mainly barren polar desert and bedrock	1976-81	190	38	158	Woo, Marsh, and Steer (1983)

Both storage and evaporation are enhanced and the runoff ratio decreases where wetland and open water conditions prevail. The "Intensive" watershed, for example, although adjacent to the "Gully River," exhibited a distinctive pattern of water balance. The "Intensive" watershed, occupied by large tracts of wetland and water, loses over half its annual precipitation to evaporation, leaving about 45 per cent available to runoff. Two rivers side-by-side and yet showing different runoff regimes demonstrate that the local environment noticeably influences the basin water balance. More hydrometeorological information thus is required before the water balance on the regional scale can be calculated reliably for the Canadian Arctic Islands.

SUMMARY AND CONCLUSIONS

Water balances for small Arctic basins are often difficult to calculate because of inadequate and inaccurate data for precipitation, runoff, evaporation and basin storage. A six-year study shows that, for a given basin, the large year-to-year variations occurring in the magnitude of the water balance's components indicate a need for a long-term data base. Regional differences in the water balance reflect spatial variation in precipitation: the magnitude of various water balance components tends to be higher for the Eastern Arctic. Runoff usually consumes the bulk of annual precipitation, a typical feature of Arctic environments.

Hydrological research in the Arctic, has advanced in the last several decades, but basic hydrometeorological data are still quite sparse. Regional studies of the water balance of the Arctic archipelago thus continue to be formidable tasks. As development progresses into the last terrestrial frontier of Canada, the demand will increase for better information on water balances. Research on the regional water balance of the Arctic has, therefore, many practical applications as well as considerable scientific merit.

ACKNOWLEDGEMENTS

I am grateful to the Polar Continental Shelf Project, Department of Energy, Mines and Resources, for a decade of logistical support, and to the Natural Sciences and Engineering Research Council for funding this research.

REFERENCES

ADDISON, P.A. (1977), Studies on evapotranspiration and energy budgets on Truelove Lowland, in L.C. Bliss (editor), *Truelove Lowland, Devon Island, Canada: A High Arctic Ecosystem* (Edmonton: University of Alberta Press): 281-300.

BOSTOCK, H.S. (1970), Physiographic subdivisions of Canada, in R.J.W. Douglas (editor), *Geology and Economic Minerals of Canada* (Ottawa: Information Canada, Economic Geology No. 1): 9-30.

BOVIS, M.J. and BARRY, R.G. (1974), A climatological analysis of northern polar desert areas, in T.L. Smiley and J.H. Zumberge (editors), *Polar Desert and Modern Man* : 23-31.

CHURCH, M. (1974), Hydrology and permafrost with reference to northern North America, *Workshop Seminar on Permafrost Hydrology, Proceedings* (Ottawa: Canadian National Committee, IHD): 7-20.

COGLEY, J.G. and McCANN, S.B. (1975), Surface runoff characteristics of an Arctic nival catchment: Mecham River, Cornwallis Island, *Canadian Hydrology Symposium, Proceedings*, 75: 282-288.

COGLEY, J.G. and MCCANN, S.B. (1976), An exceptional storm and its effects in the Canadian High Arctic, *Arctic and Alpine Research*, 8: 105-115.

COOK, F.A. (1967), Fluvial Processes in the High Arctic, *Geographical Bulletin*, 9: 262-268.

HARE, F.K. and HAY, J.E. (1971), Anomalies in the large-scale annual water balance over northern North America, *The Canadian Geographer*, 15: 79-94.

HERON, R. and WOO, M.K. (1978), Snowmelt computation for a High Arctic site, *Eastern Snow Conference, Proceedings*, 35: 162-172.

HOLECEK, G.R. and VOSAHLO, M.V. (1975), Water balance of three High Arctic river regimes, *Canadian Hydrology Symposium, Proceedings*, 75: 448-461.

JACKSON, C.I. (1960), Snowfall measurements in northern Canada, *Quarterly Journal of the Royal Meteorological Society*, 86: 273-275.

LEWKOWICZ, A.G. and FRENCH, H.M. (1982), The hydrology of small runoff plots in an area of continuous permafrost, Banks Island, N.W.T., *Canadian Permafrost Conference, Proceedings*, 4: 151-162.

LVOVITCH, M.E. (1973), The global water balance, *EOS*, 54: 28-42.

MARSH, P., ROUSE, W.R. and WOO, M.K. (1981), Evaporation at a High Arctic site, *Journal of Applied Meteorology*, 20: 714-716.

MARSH, P. and WOO, M.K. (1979), Annual water balance of small High Arctic basins, *Canadian Hydrology Symposium, Proceedings*, 79: 536-546.

MAXWELL, M.B. (1980), *The Climate of the Canadian Arctic Islands and Adjacent Waters*, Volume 1 (Ottawa: Ministry of Supply and Services): 532 pp.

McCANN, S.B. and COGLEY, J.G. (1972), Hydrologic observations on a small Arctic catchment, Devon Island, *Canadian Journal of Earth Sciences*, 9: 361-365.

OHMURA, A. (1982a), Evaporation from the surface of the Arctic tundra on Axel Heiberg Island, *Water Resources Research*, 18: 291-300.

OHMURA, A. (1982b), Regional water balance on the Arctic tundra in summer, *Water Resources Research*, 18: 301-305.

PISSART, A. (1967), Les modalites de l'écoulement de l'eau sur l'Ile Prince Patrick (76°Lat N. 120°Long. O. Artique Canadien), *Biuletyn Peryglacjalny*, 16: 217-224.

RYDEN, B.E. (1977), Hydrology of Truelove Lowland, in L.C. Bliss (editor), *Truelove Lowland, Devon Island, Canada: A High Arctic Ecosystem* (Edmonton: University of Alberta Press): 107-136.

SELLERS, W.O. (1965), *Physical Climatology* (Chicago: University of Chicago Press): 272 pp.

WALKER, E.R. and LAKE, R.A. (1975), Runoff in the Canadian Arctic Archipelago, in G. Weller and S.A. Bowlings (editors), *Climate of the Arctic* (Fairbanks: University of Alaska, Geophysical Institute): 274-378.

WEDEL, J.H., THORNE, G.A. and BARACOS, P.C. (1977), Site intensive hydrologic study of a small catchment on Bathurst Island, *Hydrological regimes, Freshwater Project No. 1* (Ottawa: Environment Canada, Inland Waters Directorate, Western and Central Region): 123 pp.

WOO, M.K. (1983), Hydrology of a drainage basin in the Canadian High Arctic, *Annals of the Association of American Geographers*: (in press).

WOO, M.K., HERON, R., MARSH, P. and STEER, P. (1983), Comparison of weather station snowfall with winter snow accumulation in High Arctic basins, *Atmospheric-Ocean*: (in press).

WOO, M.K., HERON, R. and STEER, P. (1981), Catchment hydrology of a High Arctic Lake, *Cold Regions Science and Technology*, 5: 29-41.

WOO, M.K. and MARSH, P. (1978), Analysis of error in the determination of snow storage for small High Arctic basins, *Journal of Applied Meteorology*, 17: 1537-1541.

WOO, M.K., MARSH, P. and STEER, P. (1983), Basin water balance in a continous permafrost environment, *Fourth International Permafrost Conference, Proceedings* (Fairbanks, Alaska): (in press).

WOO, M.K. and SAURIOL, J. (1980), Channel development in snow-filled valleys, Resolute, N.W.T., Canada, *Geografiska Annaler*, 62A: 37-56.

WOO, M.K. and SAURIOL, J. (1981), Effects of snow jams on fluvial activities in the High Arctic, *Physical Geography*, 2: 83-98.

WOO, M.K. and STEER, P. (1979), Measurement of trace rainfall at a High Arctic site, *Arctic*, 32: 80-84.

WOO, M.K. and STEER, P. (1982), Occurrence of surface flow on Arctic slopes, Southwestern Cornwallis Island, *Canadian Journal of Earth Sciences*, 19: 2368-2377.

3

NORTHERN DEVELOPMENT AND NATIVE PEOPLES

Robert M. Bone

"Politics and geography," remarked one of Canada's foremost historians, "face each other in an immediate and implacable confrontation" (Creighton, 1970: 2). In the unfolding of Canada's evolution from a colony to a nation, the integration of peripheral areas into the national fabric is of paramount importance. The driving force behind this territorial expansion has been the demand for resources by the North American industrial heartland. During this process of nation-building, the original inhabitants have been displaced and their principal commercial activity of trapping has been restricted.

DEVELOPMENT AND CHANGE

This relationship between economic development and native peoples forms the central theme of this paper. Since the end of the Second World War, the governments of Canada have extended a wide variety of public services into the North. These services while greatly improving the living conditions of northerners, have weakened their dependence on trapping (Usher, 1981: 416). The pace of this decline corresponds closely to the growth of urbanization of native peoples (Robinson, 1983: 322-328).

The first wave of native migrants to northern settlements in the 1950s was concerned about access to basic health services, family allowance payments, and food supplies. These migrants still regarded hunting and trapping as the basis of their economy. Their offspring, borne and raised in an urban environment by the 1970s, were much more committed to the modern way of life and were much better prepared to live and work in this environment.

The final process of frontier development likely will take place by the end of this century, and will require more readily available wage employment, especially in regional centers and mining towns (Canada, 1981: 2; Carr and Blair, 1981). At the same time, an anticipated increase in the size of the native labor force resulting from the "baby boom" of the 1960s will place additional stress on native communities. Since the demand for jobs by native youths is unlikely to be satisfied in their home communities, many either must accept a life of part-time employment and welfare or migrate to job opportunities in larger centers. A few will adopt a third option--returning to a quasi-subsistence life-style based on hunting and trapping.

This discussion examines the role of native peoples in the northern wage economy of Saskatchewan, and uses empirical data from the 1981 Census of Canada and Northern Saskatchewan Housing Needs Survey Report (1977). Three basic questions are posed: (1) what is the size and likelihood of increase in the 1980s of the native potential labor force? (2) what is the participation rate in the wage economy of native labor compared to the total national labor force? and (3) is the native participation rate affected by the spatial distribution of the northern economy?

CONCEPTUAL FRAMEWORK

A modern version of the staple theory of regional economic growth is the basis for constructing a conceptual framework here. This theory views the export product (staple) as the leading sector which sets the pace for regional economic growth. This growth affects the size and character of labor force. One frontier version of this theory introduces a modern economy into a traditional region; almost all of the initial labor force migrates from the developed core to create a "dualistic" labor market in which each group tends to be restricted to its own economy. The more dominant economy competing for a common pool of resources, namely land and labor gradually is perceived as offering a more attractive life style than that of the traditional form of livelihood. Concurrently, the aboriginal peoples shift from a land-oriented life to an urban one and are led to participate in a wage economy and to adopt an urban way of life.

This major spatial relocation of people is associated with a fundamental change in their traditional lifestyles. Attendant socio-economic shifts in turn affect the components of population growth, and in time, the age structure of society.

According to demographic transition theory, age structures are associated with degrees of development; developing countries and regions tend to have a higher proportion of their population in the younger age cohorts than that prevailing in developed countries and regions (Bone and Green, 1984).

The basic demographic changes occurring in a society shifting from a traditional way of life to a modern one result from a continuing high birth rate and a rapidly declining death rate. The structure of this population becomes very youthful and within a short period perhaps 60 to 65 per cent of its members comprise those under the age of 15 years. As this "baby boom" bulge moves through the young age cohorts into those potentially associated with the labor force, pressure is placed on the local economy to generate work.

The population of northern Canada exhibits characteristics commonly associated with developing countries. The most prevalent demographic characteristics of developing nations are very high birth rates, youthful populations, and "low" proportions of the population in the so-called productive age groups, that is between the ages of 15 and 64 years (Table 1).

Classification of population into three age groups (youth, active and old) provides an insight into the relationship between so-called productive and non-productive members of society. Such a classification utilizes the dependency ratio, in which the quotient obtained by dividing a numerator consisting of the youth and old age groups (those persons under 15 years of age and those over 64 years of age), by a denominator combining of the active group (those persons between the ages of 15 and 64 years) is multiplied by 100. This ratio tends to be high for developing nations and regions. For territories with high ratios, the immediate cost of providing public services and supplying public facilities to the non-productive members of society

TABLE 1

POPULATION BY MAJOR AGE GROUPS AND DEPENDENCY RATIO

Region	Youth (<15 Yrs.)	Active (15-64 Yrs.)	Old (>64 Yrs.)	Dependency Ratio
Africa	45	52	3	92
Latin America	39	56	4	75
North America	23	66	11	55
Canada	23	68	9	47
Northwest Territories	35	62	3	61
Census Div. 18	39	57	4	76

Source: Statistics Canada (1982); and M.D. Kent (1983).

often exceeds society's capability. Long-term costs then accrue to that society because the next generation of workers is poorly educated and trained in comparison to those in developed countries and regions. Fortunately for some developing regions, such as Canada's peripheral north, the transfer of public funds from the core region allows this poverty trap to be avoided.

THE SIZE OF THE POTENTIAL LABOUR FORCE

The northern baby boon which peaked in the 1960s has begun to affect the size of the younger cohorts of the potential labor force in Census Division 18 of northern Saskatchewan. The potential labor force has grown from about 9,700 in 1971 to nearly 14,400 in 1981 (Census of Canada) representing an annual average increase of 3.7%. Much of this population growth can be attributed to the high rate of natural increase in the 1950s and 1960s while the remainder represents the net in-migration of persons of working age. The impact of the natural rate of increase on the potential labor force is difficult to estimate because two related variables (age-standardized death rates and migration flows) are not known. However, an examination of the age cohorts in Census Division 18 for 1976 reveals the following: those

15 to 19 years of age form 21.3% of the potential labor force; those 20 to 24: 16.1%; 25 to 34: 25.6%; 35 to 44: 15.8%; 45 to 54: 12.8%; and 55 to 64: 8.4% (Bone 1983: 22). This progressive decline in the age cohorts suggests that the most powerful factor in shaping the age structure has been the rate of natural increase. Thus from 1971 to 1981, the percentage of the population between the ages of 15 and 64 years increased from 50.2% to 56.7% due to the growth of the active sector of the population and the decline in the rate of natural increase (Bone, 1983: 13). This growth corresponds well with the earlier high annual rates of natural increase: the annual rate in the early 1950s was just below 3%; from 1956 to 1971, it was well above 3% (the highest annual rate, 4%, occurred in 1965); and from 1971 to 1981, the rate declined annually and was below 2% in 1981 (Bone, 1983: 15).

This population explosion has created a very youthful age structure which is now affecting the size of the labor force. As this "bulge" moves with time through the age structure, it will affect each age group in the potential labor force (Bone, 1983: 16). The Metis of northern Saskatchewan will be affected even more strongly because their age structure has a higher proportion in the youth age group. Nearly 29% of the potential Metis labor force in 1976, for example, was in the 15 to 19 age bracket while the total population (consisting of roughly equally proportions of whites, Indians and Metis) had only 21% in the same age group (Bone, 1983: 18).

Generally speaking, the Metis tend to live in more remote communities than the white residents. The 34 communities comprising the Metis population in the Housing Needs Survey Report (Northern Municipal Council, 1977) had poor road access to other centers and to cities in southern Saskatchewan; 16% of this population did not have road access to other centers (Bone and Green, 1983: 15). These isolated and remote communities have little capacity to generate jobs and most jobs generated are part-time or seasonal ones. Their location means that there is little opportunity for residents to commute daily to employment outside their community; outside employment requires migration.

PARTICIPATION RATES IN THE WAGE ECONOMY

The participation rate is a percentage figure comparing the active labor force with the potential labor force. Statistics Canada defines the potential labor force as comprising those persons 15 years of age and older. This definition differs from the conventional one which includes only those persons between the ages of 15 years and 64 years of age. The Statistics Canada definition is used in this analysis.

Participation rates for northern Saskatchewan have traditionally been lower than those for the province. For example, in 1976, the rates were 49.2% and 62.7% respectively (Census of Canada, 1976: 6-22 to 6-34): for the Metis, the rate was 39.5% (Bone, 1983: 30). These disparities are directly related to differences between the economic base of the North and that of the province, and to the cultural fact that more native people have remained outside of the wage economy, either by choice or by circumstance. While northern nations appear to be shifting from a "bush" to a wage economy, many are believed to engage in both sectors. This mixed economy couples widespread use of "part-time" employment with the subsistence activities of hunting and trapping.

An examination of the participation rates for the 1971 and 1976 censuses provides some evidence that the wage economy is now involving more people. The degree to which this shift involves native people, however, is less clear.

The participation rates for northern Saskatchewan increased from 44.1% in 1971 to 49.2% in 1976. Five years later, this rate had probably reached 50%. This increase is associated with the expanded presence of government in the North and the growth of industry, particularly uranium mining. The provincial economy also was growing during this period. Despite the increase in northern participation rates, however, the provincial rate remained much higher. The 1971 difference of 14.1% declined by 1976 to 13.5% (discussed above) suggesting thereby that the north-south employment gap is becoming smaller. The question is thus raised, is this change a short-term fluctuation or a long-term trend?

The participation rates by age cohorts for Census Division 18 and the province of Saskatchewan are shown in Table 2. These indicate that (1) for each cohort, the provincial figures are higher than the northern ones; (2) the difference is greatest in the youngest age group (15 to 19); and (3) the participation rate increases substantially for the next age cohort (20 to 24) in both geographical areas. These northern participation rates are typical of developing regions in which the greatest increase in the potential labor force of such a young population occurs in the 15 to 19 age group. Since fewer northern than southern members in this age group are assumed to remain in high school, under equal conditions the participation rate should be higher in northern Saskatchewan. Because the opposite is true, those in the North who have left the public schooling system likely are having trouble finding work--unless they migrate to larger towns and cities. One explanation for this situation is caused in part by the very weak economic bases of northern communities which do not generate many new jobs.

SPATIAL DISTRIBUTION OF CENTERS

The urbanization of Indian, Inuit and Metis has altered their life style. Living in a settled environment inevitably leads towards more involvement in the larger Canadian economy and an acceptance of the social values and attitudes of the larger society. Stemming from these social changes is the widespread desire for a higher standard of living modeled after that perceived to exist in middle-class society of southern Canada and the United States. Fears by native leaders, such as Old Crow Chief Johnny Abel, that large-scale developments, such as the construction of a deep water port at Stokes Point on the Yukon coast, will set in motion forces which will destroy the basis of native society underscore native concern about the future (*Globe and Mail*, 8 October, 1983). This concern is not restricted to one or two chiefs but

TABLE 2

COMPARISON OF PARTICIPATION RATES BY AGE COHORTS, 1976

Region	15-19	20-24	25-34	35-44	45-54	55-64	>64	% Total Pop.
Census Div. 18	29.6	58.3	61.3	63.2	54.0	42.9	10.3	49.2
Saskatchewan	56.3	78.3	74.2	78.1	75.9	61.8	17.8	62.7
Difference	26.7	20.0	12.9	14.9	21.9	18.9	7.5	13.5

Source: Bone (1983: 29).

is deeply held by the majority of native leaders (Berger, 1977). The fear that mega projects would have a ruinous effect upon native society is based on the assumptions that (1) these industrial undertakings would harm the environment and thus diminish the natural food supply for native peoples, and (2) these projects, by offering high wages, would attract native workers into the wage economy, but upon their completion workers would be unemployed and would experience difficulty returning to a trapping economy.

Native perceptions of the process of northern development have substance. Because the forces driving this development already have been released, however, the character of northern Canada and the place of native peoples in their homeland are changing. How will the native people adapt to these changes?

Evidence from the socio-economic data compiled for the Metis of northern Saskatchewan in the Housing Needs Survey Report (Northern Municipal Council, 1977) sheds some light on this question. It reveals that wage earnings are important in both accessible and remote communities because many basic needs can be satisfied only by cash purchase. Distribution of wages, however, differs in the two groups of settlements. The distribution of wage earnings in accessible centers indicates a larger "middle" income group and a stronger dependence on the wage economy by most members. The wage distribution in remote centers appears related to a strategy whereby some earnings finance a more desirable activity--hunting and trapping.

CONCLUSIONS: PROSPECTS FOR ACCOMMODATION

As many natives wish to continue a life-style based on hunting and trapping, they face the problem of how to obtain a cash income which will provide the basic needs associated with the modern society. The solution may lie more in the model of a guaranteed annual income rather than in wage employment because the economies of remote settlements do not have the capacity to generate sufficient cash income to allow all local residents to satisfy their basic needs. Such a model of a basic life-style within a modern economy is occurring in northern Quebec. Here, the outcome of the James Bay and northern Quebec land claims agreement provides for a guaranteed annual income to Cree hunters and trappers. Both groups live in settlements, although the hunters and their families must spend part of the year on the land. Since the inception of this program, about half of the population of these Crees has enrolled while the other half has opted for a future based on wage employment (LaRusic, 1983: 5).

Future pressure on native peoples to choose a modern life-style will likely result in many having to move to centers where employment opportunities exist. This process of adjustment will have an impact on many small native centers which like small prairie towns in the 1950s, will disappear. Others may survive with tiny groups of traditionalists wishing to pursue relatively uncomplicated life-styles based on hunting and trapping. The concentration of most native peoples into fewer and fewer urban places ironically should ease the pressure on wildlife and thereby enhance the ability of those remaining in small centers to obtain sufficient supplies of wild meat. Such a North will likely occur only after native land claims are settled to provide financial support for the traditional economy along the lines of those contained in the James Bay and northern Quebec agreements.

REFERENCES

BERGER, T.R. (1977), *Northern Frontier, Northern Homeland. The Report of the Mackenzie Valley Pipeline Inquiry,* I, II (Ottawa: Supply and Services).

BONE, R.M. (1972), The population of Northern Canada, in W.C. Wonders (editor), *The North* (Toronto: University of Toronto Press): 91-116.

BONE, R.M. (1983), *The Impact of Wage Employment Upon Selected Metis in Northern Saskatchewan* (Saskatoon: Progress Report for the Social Sciences and Humanities Research Council), Typed manuscript.

BONE, R.M. and Green M.B. (1984), The northern native labor force: A disadvantaged work force, *The Operational Geographer,* (3).

BONE, R.M. and Green, M.B. (1983), Jobs and access - A northern dilemma, *Journal of Canadian Studies,* (Fall).

CANADA (1981), *Economic Development for Canada in the 1980s* (Ottawa: Supply and Services).

CARR, S.G.E. and BLAIR,S.R. (1981), *A Report by the Major Projects Task Force on Major Capital Projects in Canada to the Year 2000* (place and publisher not identified).

CRIEGHTON, D. (1970), *Canada's First Century* (Toronto: Macmillan).

GLOBE AND MAIL, 8 October, 1983.

KENT, M.M. (1983), *1983 World Population Data Sheet* (Washington: Population Reference Buréau).

LARUSIC, I.E. (1983), The income security program for Cree hunters and trappers: Population issues, *Proceedings: Northern Population Workshop IV* (Montreal: University of Montreal, Committee on Northern Population Research): 1-7.

NORTHERN MUNICIPAL COUNCIL (1977), *Northern Saskatchewan Housing Needs Survey Report, 1976-1977* (Saskatoon: University of Saskatchewan, Institute for Northern Studies).

ROBINSON, J.L. (1983), *Concepts and Themes in the Regional Geography of Canada* (Vancouver: Talonbooks).

STATISTICS CANADA (1982), *1981 Census of Canada: Population, Occupied private Dwellings, Private Households, Census Families in Private Households, Saskatchewan* (Ottawa: Supply and Services).

USHER, P.J. (1981), in L.D. McCann (editor), *A Geography of Canada: Heartland and Hinterland* (Scarborough: Prentice-Hall): 410-456.

WONDERS, W.C. (1960), Postwar settlement and trends in the Mackenzie Valley area, *Geografiska Annaler*, 42(4): 333-338.

Geography Faculty and some Graduate Students, U.B.C., 1954

4

PROBLEMS OF WATER MANAGEMENT IN CANADA

Bruce Mitchell

Although Canada has only one per cent of the world's population and nearly nine per cent of global runoff, water management issues represent a significant item on the agenda associated with public policy (Foster and Sewell, 1981). Southern Alberta and southern Saskatchewan supplement inadequate natural water supplies for farming through extensive irrigation systems. Winnipeg, in contrast has constructed a floodway to divert excess flood waters around the city. Deterioration of water quality in the Great Lakes and the St. Lawrence River has been a long-standing problem, while dangers from acid rain have become apparent in Ontario and Quebec only since the late 1970's. The implications for Canada of the Garrison and Ross Dam projects in the United States have received attention from media across the country.

The significance of water as a mechanism for and constraint upon development has been recognized for many years (Kristjanson and Sewell, 1961; Quinn, 1977; Day, 1983). Research contributions and needs have been reviewed both in general terms (Bruce and Maasland, 1968; Science Council of Canada, 1968) and with reference to geographical work (Sewell and Burton, 1967; Sewell, 1968; Sewell and Mitchell, 1981). This paper addresses the changing nature of water management issues and the responses found in research by geographers, including analyses of various water issues in different regions of the country. Such issues include flood plain management, water supply, drainage, water quality, navigation and transportation, and river basin management. The paper concludes by suggesting opportunities for future research.

FLOOD PLAIN MANAGEMENT

As many as 200 communities across Canada have potential flooding problems (Environment Canada, 1983: 41). Although floods often are localized and cause only minor damage, floods with substantial associated damages have occurred in Vancouver (1948), Winnipeg (1950), Toronto (1954) and Fredericton (1974). Following the Fraser River flood in 1948,

$22 million was paid in compensation; after the Red River flood of 1950, some $25 million was paid. Nearly $25 million also was paid in Ontario following Hurricane Hazel in 1954. By 1970, immediate flood damage assistance across the country had climbed past $100 million (Environment Canada, 1975: 203). During 1974, 1976 and 1979 floods caused extensive damage. Floods during 1974, in six provinces and the Yukon Territory, caused damages of $70 million. The Montreal Region in 1976 itself suffered about $30 million in flood damages. Clearly, flooding has created major problems throughout the country.

Flood problems probably have received by geographers the longest and most sustained research of any issues of water management. Burton (1965) provided an initial survey of flood damage potential on a country-wide scale. Sewell (1965) conducted a detailed analysis of the flood problem and range of adaptive strategies in the Fraser River watershed. Mitchell et al. (1978) produced a similarly detailed assessment for the Grand River valley in Ontario. Experiences in the Grand and Credit River watersheds subsequently were compared (Gardner and Mitchell, 1980). Barker and Morgan (1981) have evaluated shore and floodplain management in British Columbia and the state of Washington.

All of this research effort has been influenced by White's (1973) innovative approach to analysis of natural hazards. The conclusions from such research have pointed toward the need for a mix of structural and non-structural adjustments. A major step in that direction was taken during 1975 when the federal government introduced its Flood Damage Reduction Program which explicitly called for both structural and non-structural responses (MacLock and Page, 1980; Page, 1980).

Having advocated greater use of non-structural measures geographers now need to examine some of the problems associated with their implementation. The economic and social impact of designating flood-prone areas has created uncertainty about property values and development opportunities (Babcock and Mitchell, 1980). Conflict between flood protection and preservation of heritage buildings also has emerged (Bennett and Mitchell, 1983). The utility of flood risk maps needs to be evaluated (Handmer, 1980).

Opportunities exist to examine related problems such as soil erosion (Mailloux and Dubé, 1959) and degradation of river banks and ravines (Dufour and Lemieux, 1978). A related problem receiving recent attention is the management of coastal and shoreline hazards. Nelson et al. (1975) and Day (1977) have led an extensive program of research on this aspect along the north shore of Lake Erie (Fraser et al., 1977; Needham and Nelson, 1977; Kreutzwiser, 1982). The Canadian Council of Resources and Environment Ministers (1979) has reviewed shore zone management problems across the nation.

Despite the considerable research devoted to flood-related problems, much more investigation is needed, especially to integrate our understanding of both natural and human systems, and to recognize regional and cultural differences across the country in the design of adaptive measures. Non-structural measures also must be studied to avoid or resolve problems affecting their implementation.

DRAINAGE

Research on flooding has primarily focused urban areas. Excess water in rural areas in contrast, has led to research on drainage policy and practice, especially in Ontario where many inter-related aspects have drawn attention. On one hand, drainage programs must be examined in conjunction with agricultural policies designed to increase food production. On the other hand, drainage activity must be coordinated with policies to maintain and enhance wetland areas which are valued for their role in controlling flood discharges and in providing a habitat for wildlife, especially migratory birds.

Several geographers (Kettel and Day, 1974; Found, Hill and Spence, 1974; Found and Spence, 1977) have conducted research which was utilized by the Ontario Select Committee on Land Drainage (1974) during its review of the province's land drainage policy and practice. These investigations focused upon the economic and environmental benefits and costs associated with land drainage.

The Ontario Select Committee commissioned Found, Hill and Spence (1974) to examine an Ontario sample of 37 drainage projects distributed among 7 townships. The majority of drainage projects had benefit-cost ratios in excess of 1.0 after operation commenced. A disconcerting finding, nevertheless, was that, assuming a 20-year life span for drains and an interest rate of 6 per cent, some 11 of 37 drainage projects had ratios less than 1.0. These findings, and those for small reservoirs reported by Day (1974; 1977), suggest that preconstruction benefit-cost ratios often have been inaccurate guides for policy makers. Considerable opportunity exists, therefore, to refine existing methods of pre-construction evaluation and/or to develop alternative methods pertinent to drainage and other water development activities.

WATER SUPPLY

With nine per cent of the world's total river flow, Canada is unlikely to suffer an overall shortage of water. Nevertheless, because water doe not fall at the places or times of greatest need, redistribution will be required to alleviate local shortages.

Water is supplied from two sources. About ninety per cent of water used in Canada comes from surface sources. The other ten per cent is groundwater supplied either through municipal water systems or from private wells. Groundwater is a significant source of domestic water in rural areas, and also is an important source for industrial and irrigation water. Kitchener-Waterloo in Ontario is alone among the 22 standard metropolitan areas in Canada in being totally dependent upon groundwater.

Using data assembled by Environment Canada, Foster and Sewell (1981: 17) have drawn attention to six regions of the country in which current and future demands either exceed or are within 20 per cent of total available monthly flow. For several of these areas, annual reliable flow cannot satisfy present or estimated demands. As a result, serious problems of water supply are anticipated in basins associated with the (1) Okanagan, (2) Milk, (3) North Saskatchewan, (4) South

Saskatchewan, (5) Red-Assiniboine, and (6) Southern Ontario. Against such a background, research by geographers can be viewed.

Development of urban water supplies has been examined by geographers. Guyer (1971) reviewed the decision process which resulted in London, Ontario constructing a pipeline from Lake Huron. MacIver (1970) analyzed the way in which the range of urban water supply alternatives was perceived and explored by various interests in the Grand River valley of southwestern Ontario. He demonstrated clearly the major problems which water managers can encounter when the values of the ultimate water users are not incorporated during the search for sources of water. Conflict of a different nature was addressed by Wood (1976) during an examination of how the Goldstream River on Vancouver Island was managed to service water needs of Victoria and to accommodate the requirements of spawning salmon.

The need to redistribute water resources to match supply against demand has led to several inquiries focused upon water diversions. A fine review of water diversion experience in Canada is by Laycock and Reinelt (1981). Quinn (1971; 1981), maintaining an up-to-date inventory of interbasin transfers across Canada, has identified 60 water transfer projects which either were functioning or under construction in 1980 (Quinn, 1981: 68-69). Of all agents, hydropower generation dominated the number and scale of transfers. Other uses included irrigation, flood control and municipal water supply. Evaluations of several diversions after they began operating have been completed in northwestern Ontario (Peet and Day, 1980; Day _et al._, 1982).

Given the shortages of natural water supplies in the southern Prairies, and the consequent importance of irrigation, the paucity of related geographical research is surprising (Canada, 1982). Nearly 60 percent of Canada's irrigated acreage is found on the prairies (Environment Canada, 1975: 121). Irrigation and stock watering on the prairies, for example, account for just under 70 per cent of total Canadian agricultural water withdrawls. Despite the work by Raby (1965) on irrigation development in Alberta, and by McKnight (1979) on the use of centre pivot irrigation on the prairies, irrigation is a subject requiring much more attention.

An area, however, pursued by some geographers involves the interrelation of forecasting-pricing mechanisms and demand management. Tate and Lacelle (1978) have reviewed municipal water use across Canada, while Mitchell and Leighton (1977) have developed a forecasting technique for smaller communities. Grima (1972; 1973) and Sewell and Roueche (1974) have analyzed the impact of pricing strategies on water use and demand. As managers come increasingly to consider water demand as well as the more traditional aspects of supply, more opportunities are arising for applied geographical research.

Water supply research thus has covered numerous aspects of municipal supplies, interbasin transfers, and demand management. While they all deserve continued effort, research on various aspects of irrigation has been gravely neglected by geographers and hopefully will be addressed more vigorously in the future.

WATER QUALITY

Water quality is probably the most significant current national issue in Canadian water management. Problems range from pollution of local water courses by municipal sewage treatment systems, to the disposal of toxic wastes, including nuclear materials, the deterioration of the Great Lakes basin, and the emergence of acid rain as a pervasive problem.

Against this background of enduring and arising issues for management, only a few geographers have explicitly addressed water quality in their research. One problem to receive attention is the way in which perception of water quality affects recreational behavior. Barker (1971), focusing upon the criteria and attitudes of managers and the general public in evaluating the impact of water pollution upon water-based recreation, utilized two questionnaire surveys on twelve beaches in southern Ontario and in metropolitan Toronto. Parkes (1973; 1974) studied the public's perception of water quality, the impact of water quality on recreation, and the willingness of the public to pay for improvement. He selected seven high-use water-oriented sites in three provinces (Nova Scotia, Quebec, Saskatchewan) and completed over 1,000 interviews. Given the importance of water-based recreation in Canada, this area of research needs further development.

Alternative regulations to control water pollution have been reviewed by Grima and Griffith (1983). Grima also has been active in analyzing the role of public participation in the evolution of strategies to improve water quality in the Great Lakes (Grima, 1980; Grima and Wilson-Hodges, 1977; Grima and Mason, 1983). The field of public participation in water management indeed is one of which Canadian geographers have been active (O'Riordan, 1976; Sewell and O'Riordan, 1976; Griffith, 1979; Mitchell, 1983). A need exists to explore alternative methods to facilitate public participation which will keep the decision process accessible to the public but which will also streamline the time required to make decisions.

Three issues appear to dominate the foci in the 1980s of research into water quality. The consequences of acid rain are both grave and widespread (Ontario, 1980; Canada, 1981). Disposal of toxic wastes represents a problem of ever-increasing importance especially in the search for a balance between economic development and environmental quality. Disposal of toxic wastes fits into the more general category of "noxious facilities" research which has been well developed by geographers and which draws on the tradition of spatial analysis in geography. More such work would provide a counterbalance to the predominance in Canadian water-oriented geographical research of the man-environment tradition. Finally, given the importance of the Great Lakes region in the Canadian economy, concern for the quality of water in those lakes should continue to sustain research by geographers and others (Pleva, 1967).

NAVIGATION AND TRANSPORTATION

The rivers of Canada were the early highways allowing explorers to penetrate the heart of the continent. Subsequently, the rivers and lakes provided the means by which furs, logs and other resource products were moved to coastal ports for export. Transportation thus represents one of the earliest continuous uses of Canadian water.

As a result, the use of interior Canadian waters for navigation and transport is an obvious aspect for geographical research, especially because it focuses upon spatial interaction. Relatively little has been done, although commercial use of water bodies such as the St. Lawrence Seaway has generated the greatest amount of research. Camu (1949; 1956; 1957) has examined traffic patterns and toll schedules on the seaway, while Kerr (1964) has analyzed its role in trade on the Great Lakes. Lasserre (1967; 1973; 1980) and Wood (1956) focused upon some of the impacts of the seaway on adjacent cities and regions. Some of their early work, if done today, would be labeled, "social impact assessment."

Recreational use of waterways also has attracted research interest. Helleiner's (1981) graph-theoretic analysis of recreational boat trips on the Trent-Severn Waterway in eastern Ontario represents one of several important types of geographic research focused upon water-based recreation (Jacobs and Williams, 1982). Another, which is likely to draw increasing attention, pertains to the development of underwater parks along the shorelines of various Canadian regions (Marsh and Huff, 1982). Nevertheless, whether their focus is upon recreational or commercial uses of water for transport and navigation, many more geographical inquiries need to be conducted in this field.

RIVER BASIN MANAGEMENT

Integrating the regional contexts of water and associated land-based resources provides a superb testing ground for the synthesizing skills of the geographer. Tate (1981) has provided a fine overview of river basin management in Canada, while Mitchell and Gardner (1983) have edited a monograph containing assessments of specific experiences across the country. O'Riordan and O'Riordan (1972) have given a detailed assessment of the Okanagan river basin planning exercise, and Mathewson (1981) has described planning procedures being used in Ontario.

At the international scale, Chapman (1963) edited a volume focused upon a mix of experiences in different countries. Friesen (1979) and

Friesen and Day (1977) have completed a careful analysis of the power-sharing agreement between Canada and the United States on the Niagara River. Sewell (1966) reviewed the implications of the Columbia River Treaty. Ross and Marts (1975) studied the implications of the High Ross Dam in Washington State for the Skagit Valley of British Columbia.

The design of institutional arrangements for river basin management has also generated considerable research activity. Raby (1964) investigated the role of Alberta in the Prairie Provinces Water Board, an organization involving the three Prairie provinces. Pleva (1961a; 1961b) reviewed the strengths and weaknesses of the Ontario Conservation Authorities. That work was extended by Mitchell (1983), who previously had considered approaches to conservation in other regions of Canada (Mitchell, 1975). Carlyle (1980; 1983) has done comparable work for the Conservation Districts of Manitoba. These studies suggest that if administrative areal units are to be compatible with natural spatial units like the watershed, then further research is needed into institutional arrangements.

CONCLUSIONS: PATTERNS AND OPPORTUNITIES

Problems of water management have evolved over time to affect most Canadian regions. Water supply and flood control were the predominant concerns immediately after World War II. Those problems continue to be important, usually in highly localized situations. Deterioration of water quality has emerged as the current major water problem. The effects of acid rain appearing in the late 1970s have dramatized how widespread and difficult that problem has become. Coastal or shoreline management emerged as a concern during the mid 1970s, and like comprehensive river basin management continues to draw research interest. The fields of irrigation, drainage, transportation and navigation, however, have been relatively neglected.

Many opportunities exist for research. Data needs must be more systematically considered (Quinn and Spence, 1976; Church and Kellerhals,

1980). Although the skills of physical geographers are pertinent to the processes of management, more physically-based studies need to be conceived with the needs of management in mind (Foster, 1972; Rasid, 1979). Careful evaluations of ongoing exercises in management can be used to help adjust existing activities or to improve future decisions (Day, 1974; 1977). Abundant opportunities for research thus exist which extend from the local to the international scale, from the west coast to the Atlantic shores, and from the perspective of the historical geographer to that of the geomorphologist.

REFERENCES

BABCOCK, M. and MITCHELL, B. (1980), Impact of flood hazard on residential property values in Galt (Cambridge), Ontario, *Water Resources Bulletin*, 16(3): 532-537.

BARKER, M.L. (1971), Beach pollution in the Toronto region, in W.R.D. Sewell and I. Burton (editors), *Perceptions and Attitudes in Resources Management* (Ottawa: Information Canada): 37-47.

BARKER, M.L. and MORGAN, W.B. (1981), Shore and floodplain management strategies in British Columbia and Washington State: An international comparison, *Canadian Water Resources Journal*, 6(1): 3-24.

BENNETT, W.J. and MITCHELL, B. (1983), Floodplain management: Land acquisition versus preservation of historic buildings in Cambridge, Ontario, Canada, *Environmental Management*, 7(4): 327-337.

BRUCE, J.P. and MAASLAND, D.E.L. (1968), *Water Resources Research in Canada* (Ottawa: Queen's Printer).

BURTON, I. (1965), Flood-damage reduction in Canada, *Geographical Bulletin*, 7(3): 161-186.

CAMU, P. (1949), The traffic on the Upper St. Lawrence River, *Revue Canadienne de Géographie*, 7: 28-30.

CAMU, P. (1957), Tolls on the St. Lawrence Seaway, *The Canadian Geographer*, 10: 1-12.

CAMU, P. (1957), Tolls on the St. Lawrence Seaway, *The Canadian Geographer,* 10: 1-12.

CANADA, PRAIRIE FARM REHABILITATION ADMINISTRATION (1982), *History of Irrigation in Western Canada* (Regina: Prairie Farm Rehabilitation Administration).

CANADA SUB-COMMITTEE ON ACID RAIN OF THE STANDING COMMITTEE ON FISHERIES AND FORESTRY (1981), *Still Waters* (Ottawa: Minister of Supply and Services).

CANADIAN COUNCIL OF RESOURCE AND ENVIRONMENT MINISTERS (1979), *Shore Zone Management Symposium: Proceedings* (Toronto: Canadian Council of Resource and Environment Ministers).

CARLYLE, W.J. (1980), The management of environmental problems on the Manitoba Escarpment, *The Canadian Geographer,* 24(3): 255-269.

CARLYLE , W.J. (1983), Agricultural drainage in Manitoba: The search for administrative boundaries, in B. Mitchell and J.S. Gardner (editors), *River Basin Management: Canadian Experiences* (Waterloo: University of Waterloo, Department of Geography Publication Series No. 20): 279-295.

CHAPMAN, J.D. (editor) (1963), *The International River Basin: Proceedings of a Seminar on the Development and Administration of the International River Basin* (Vancouver: University of British Columbia Publications Centre).

CHURCH, M. and KELLERHALS, R. (1980), On hydrological data needs in Canada, *Canadian Water Resources Journal,* 5(4): 1-24.

DAY, J.C. (1974), Benefit-cost analysis and multiple-purpose reservoirs: A reassessment of the Conservation Authorities Branch Deer Creek Project, Ontario, in F. Leversedge (editor), *Priorities in Water Management* (Victoria: University of Victoria, Western Geographical Series Vol. 8): 23-34.

DAY, J.C. (1977), The Parkhill dam and reservoir, in R.R. Krueger and B. Mitchell (editors), *Managing Canada's Renewable Resources* (Toronto: Methuen): 308-323.

DAY, J.C. (1983), *Major Canadian Water Management Issues,* Paper prepared for the China-Canada Conference (Hamilton: McMaster University, Department of Geography).

DAY, J.C., FRASER, J.A. and KREUTZWISER, R.D. (1977), Assessment of flood and erosion assistance programs, Rondeau coastal experiences, Lake Erie, *Journal of Great Lakes Research*, 3(1-2): 38-45.

DAY, J.C., BRIDGER, K.C., PEET, S.E. and FRIESEN, B.F. (1982), North-western Ontario river diversions, *Water Resources Bulletin*, 18(2): 27-306.

DUFOUR, J. and LEMIEUX, G.H. (1978), L'aménagement des berges, ravins et monts urbains dans la conurbation du Haut-Saguenay: vers un espace fonctionnel, *Cahiers de Géographie du Québec*, 22(57): 421-435.

ENVIRONMENT CANADA, (1975), *Canada Water Year Book, 1975* (Ottawa: Information Canada).

ENVIRONMENT CANADA (1983), *Canada Water Year Book, 1981-1982* (Ottawa: Minister of Supply and Services).

FISHERIES AND ENVIRONMENT CANADA (1976), *Canada Water Year Book, 1976* (Ottawa: Minister of Supply and Services).

FOSTER, H.D. (1972), Geomorphology and water resource management: Portage Inlet, a case study on Vancouver Island, *The Canadian Geographer*, 16(2): 128-143.

FOSTER, H.D. and SEWELL, W.R.D. (1981), *Water: the Emerging Crisis in Canada* (Ottawa: Canadian Institute for Economic Policy).

FOUND, W.C., HILL, A.R. and SPENCE, E.S. (1974), *Economic and Environmental Impacts of Land Drainage in Ontario* (Toronto: York University, Department of Geography Monograph No. 4).

FOUND, W.C. and SPENCE, E.S. (1977), The Drainage Act, 1975: Its evolution and possible impacts, *Canadian Water Resources Journal*, 2(2): 51-61.

FRASER, J.A., DAY, J.C., KREUTZWISER, R.D. and TURKHEIM, R.J. (1977), Residents' utilization of coastal hazard assistance programs, the Long Point Area, Lake Erie, *Canadian Water Resources Journal*, 2(2): 37-50.

FRIESEN, B.F. (1979), The international sharing of Niagara River hydroelectric power diversions, *Canadian Water Resources Journal*, 4(4): 26-38.

FRIESEN, B.F. and DAY, J.C. (1977), Hydroelectric power and scenic provisions of the 1950 Niagara Treaty, *Water Resources Bulletin*, 16(6): 1175-1189.

GARDNER, J.S. and MITCHELL, B. (1980), Floodplain management in Ontario: An analysis of existing and proposed policy in the Grand and Credit River watersheds, *Journal of Environmental Management*, 11(2): 119-132.

GRIFFITH, C. (1979), Water development and public involvement: A case study of the proposed Glengowan Dam and Reservoir, *Canadian Water Resources Journal*, 4(2): 72-81.

GRIMA, A.P. (1972), *Residential Water Demand: Alternative Choices for Management* (Toronto: University of Toronto Press).

GRIMA, A.P. (1973), The impact of policy variables on residential water demand and related investment requirements, *Water Resources Bulletin*, 9(4): 703-710.

GRIMA, A.P. (1980), The utilization of public input in water quality management: Case study and discussion, in W.R.D. Sewell and M.L. Barker (editors), *Water Problems and Policies*, Cornett Occasional Paper No. 1 (Victoria: University of Victoria, Department of Geography): 173-186.

GRIMA, A.P. and WILSON-HODGES, C. (1977), Regulation of Great Lakes water levels: The public speaks out, *Journal of Great Lakes Research*, 3(6): 240-257.

GRIMA, A.P. and GRIFFITH, C.R. (1983), Complementary approaches to regulation in water quality management, in B. Mitchell and J.S. Gardner (editors), *River Basin Management: Canadian Experiences* (Waterloo: University of Waterloo, Department of Geography Publication Series No. 20): 335-352.

GRIMA, A.P. and MASON R.J. (1983), Apples and oranges: Toward a critique of public participation in Great Lakes decision, *Canadian Water Resources Journal*, 8(1): 22-50.

GUYER, A. (1971), Water supply for London, Ontario, in W.R.D. Sewell and I. Burton (editors), *Perceptions and Attitudes in Resources Management* (Ottawa: Information Canada), 61-69.

HANDMER, J. (1980), Flood hazard maps as public information: An assessment within the context of the Canadian Flood Damage Reduction Program, *Canadian Water Resources Journal*, 5(4): 82-100.

HELLEINER, F.M. (1981), The regionalization of a waterway: A study of recreational boat traffic, *The Canadian Geographer*, 25(1): 60-74.

JACOBS, H. and WILLIAMS, P. (1982), Recurring policy issues in water-based recreation management, *Canadian Water Resources Journal*, 7(3): 1-10.

KERR, D. (1964), The St. Lawrence Seaway and trade on the Great Lakes, 1958-63, *The Canadian Geographer*, 8(4): 188-196.

KETTEL, G.A. and DAY, J.C. (1974), Outlet drainage in Ontario - a methodological exploration, *Journal of Environmental Management*, 2(4): 331-350.

KREUTZWISER, R. (1982), An evaluation of government response to Lake Erie shoreline flood and erosion hazard, *The Canadian Geographer*, 26(3): 263-273.

KRISTJANSON, K. and SEWELL, W.R.D. (1961), Water management problems and issues in Canada, *Resources for Tomorrow: Conference Background Papers*, Vol. 1 (Ottawa: Queen's Printer): 205-210.

LASSERRE, J.-C. (1967), Le rôle de l aménagement du Saint-Laurent dans l'évolution de la cité de Cornwall (Ontario), *Cahiers de Géographie du Québec*, 11(23): 307-325.

LASSERRE, J.-C. (1973), Le complexe portuaie de la Côte-Nord et son rôle dans la navigation sur le Saint-Laurent, *Cahiers de Géographie du Québec*, 17(40): 155-169.

LASSERRE, J.-C. (1980), La voie maritime du Saint-Laurent: atout ou handicap pour le Québec? *Cahiers de Géographie du Québec*, 24(63): 465-474.

LAYCOCK, A.H. and REINELT, B.R. (editors) (1981), Special Issues on Interbasin Transfers, *Canadian Water Resources Journal*, 6(1): 64-121; 6(2): 1-105.

MACIVER, I. (1970), *Urban Water Supply Alternatives: Perception and Choice in the Grand Basin Ontario* (Chicago: University of Chicago, Department of Geography Research Paper No. 126).

MACLOCK, R.B. and PAGE, G.A. (1980), Cutting our flood losses, in W.R.D. Sewell and M.L. Barker (editors), *Water Problems and Policies,* Cornett Occasional Paper No. 1 (Victoria: University of Victoria, Department of Geography): 7-12.

MAILLOUX, A. and DUBE, A. (1959), Erosion et conservation des sols dans la région du Lac-Saint-Jean-Saguenay, *Cahiers de Géographie du Québec,* 3(5): 77-84.

MARSH, J.S. and HUFF, D. (1982), Underwater parks: Possibilities, problems and Canadian initiatives, *Canadian Water Resources Journal,* 7(3): 69-71.

MATHEWSON, S.A. (1981), Watershed management, *Watershed Planning and Management* (Thornhill: Ontario Chapter, Soil Conservation Society of America): 3-8.

McKNIGHT, T.L. (1979), Centre pivot irrigation: The Canadian experience, *The Canadian Geographer,* 23(4): 360-367.

MITCHELL, B. (1975), *Institutional Arrangements for Water Management: Canadian Experiences* (Waterloo: University of Waterloo, Department of Geography Publication Series No. 5).

MITCHELL, B. (1983), River basin planning: Problems, opportunities and implications for Ontario Conservation Authorities, *Canadian Water Resources Journal,* 8(1): 1-21.

MITCHELL, B. and LEIGHTON, B. (1977), A comparison of multivariate and trend forecasting estimates with actual water use, *Water Resources Bulletin,* 13(4): 817-824.

MITCHELL, B., GARDNER, J., COOK, R. and VEALE, B. (1978), *Physical and Institutional Arrangements for the Urban Flood Hazard: Grand River Watershed* (Waterloo: University of Waterloo, Department of Geography Publication Series No. 13).

MITCHELL, B. and GARDNER, J.S., (editors) (1983), *River Basin Management: Canadian Experiences* (Waterloo: University of Waterloo, Department of Geography Publication Series No. 20).

NEEDHAM, R.D. and NELSON, J.G. (1977), Newspaper response to flood and erosion hazards on the North Lake Erie shore, *Environment Management*, 10(6): 521-540.

NELSON, J.G., BATTIN, J.G., BEATTY, R.A. and KREUTZWISER, R.D. (1975), The Fall 1972 Lake Erie floods and their significance to resources management, *The Canadian Geographer*, 19(1): 35-59.

ONTARIO, Ministry of the Environment (1980), *The Case Against Acid Rain* (Toronto: Ministry of the Environment).

ONTARIO, Select Committee on Land Drainage (1974), *Agricultural Land Drainage in Ontario* (Toronto: Ministry of Agriculture and Food).

O'RIORDAN, J. (1976), The public involvement program in the Okanagan Basin study, *Natural Resources Journal*, 16(1): 177-196.

O'RIORDAN, T. and O'RIORDAN, J. (1972), *Okanagan Water Decisions* (Victoria: University of Victoria, Western Geographical Series Vol. 4).

PAGE, G.A. (1980), The Canadian Flood Damage Reduction Program, *Disasters*, 4(4): 411-421.

PARKES, J.G.M. (1973), *Public Perceptions of Water Quality and Their Effect on Water-Based Recreation* (Ottawa: Information Canada).

PARKES, J.G.M. (1974), User response to water quality and water-based recreation in the Qu'Appelle Valley, Saskatchewan, in F. Leversedge (editor), *Priorities in Water Management* (Victoria: University of Victoria, Western Geographical Series Vol. 8): 99-112.

PEET, S.E. and DAY, J.C., (1980), The Long Lake Diversion: An environmental evaluation, *Canadian Water Resources Journal*, 5(3): 34-48.

PLEVA, E.G. (1961a), Multiple purpose land and water districts in Ontario, in H. Jarrett (editor), *Comparisons in Resource Management* (Baltimore: John Hopkins): 189-207.

PLEVA, E.G. (1961b), The Thames River Valley, *Resources for Tomorrow: Conference Background Papers*, Vol. 1 (Ottawa: Queen's Printer): 471-477.

PLEVA, E.G. (1967), Environmental control in the Great Lakes-St. Lawrence Region of North America, in C.E. Dolman (editor), *Water Resources of Canada* (Toronto: University of Toronto Press): 83-86.

QUINN, F. (1971), Continental water images: Past and present, in W.R.D. Sewell and I. Burton (editors), *Perceptions and Attitudes in Resources Management* (Ottawa: Information Canada): 81-87.

QUINN, F. (1977), Notes for a National Water Policy, in R.R. Krueger and B. Mitchell (editors), *Managing Canada's Renewable Resources* (Toronto: Methuen): 226-238.

QUINN, F. (1981), Water transfers - Canadian style, *Canadian Water Resources Journal*, 6(1): 64-76.

QUINN, F.J. and SPENCE, E.S. (1976), Canada's surface water resources: inventory and management, in G.R. McBoyle and E. Sommerville (editors), *Canada's Natural Environment: Essays in Applied Geography* (Toronto: Methuen): 72-107.

RABY, S. (1964), Alberta and the Prairie provinces Water Board, *The Canadian Geographer*, 8(2): 85-91.

RABY, S. (1965), Irrigation development in Alberta, *The Canadian Geographer*, 9(2): 31-40.

RASID, H. (1979), The effects of regime regulation by the Gardiner Dam on downstream geomorphic processes in the South Saskatchewan River, *The Canadian Geographer*, 23(2), 140-158.

ROSS, W.M. and MARTS, M.E. (1975), The High Ross Dam project: Environmental decisions and changing environmental attitudes, *The Canadian Geographer*, 19(3): 221-234.

SCIENCE COUNCIL OF CANADA (1968), *A Major Program of Water Resources Research in Canada*, Science Council of Canada Report No. 3 (Ottawa: Queen's Printer).

SEWELL, W.R.D. (1962), Multiple-purpose development of Canada's water resources, *Water Power*, 14(April): 146-151.

SEWELL, W.R.D. (1965), *Water Management and Floods in the Fraser River Basin* (Chicago: University of Chicago, Department of Geography Research Paper No. 100).

SEWELL, W.R.D. (1966), The Columbia River treaty: Some lessons and implications, *The Canadian Geographer*, 10(3): 145-156.

SEWELL, W.R.D. (1968), The contribution of social science research to water resource management in Canada, in J.P. Bruce and D.E.L. Maasland (editors), *Water Resources Research in Canada* (Ottawa: Queen's Printer): 117-160.

SEWELL, W.R.D. (1969), Geographical research in water management in Canada: Inventory and prospect, in W.R.D. Sewell, R.W. Judy and L. Ouellet (editors), *Water Management Research: Social Science Priorities* (Ottawa: Queen's Printer): 57-82.

SEWELL, W.R.D. and BURTON, I. (1967), Resource development policy, *The Canadian Geographer*, 11(4): 327-340.

SEWELL, W.R.D. and ROUECHE, L. (1974), Peak load pricing and urban water management: Victoria, B.C., a case study, *Natural Resources Journal*, 14(3): 383-400.

SEWELL, W.R.D. and O'RIORDAN, T. (1976), The culture of particip-ation in environmental decision-making, *Natural Resources Journal*, 16(1): 1-21.

SEWELL, W.R.D. and MITCHELL, B. (1981), The way ahead, in B. Mitchell and W.R.D. Sewell (editors), *Canadian Resource Policies: Problems and Prospects* (Toronto: Methuen): 262-284.

TATE, D. (1981), River basin development in Canada, in B. Mitchell and W.R.D. Sewell (editors), *Canadian Resource Policies: Problems and Prospects* (Toronto: Methuen): 151-179.

TATE, D. and LACELLE, D. (1978), Municipal water use in Canada, *Canadian Water Resources Journal*, 3(2): 61-78.

WHITE, G.F. (1973), Natural hazards research, in R.J. Chorley (editor), *Directions in Geography* (London: Methuen): 193-216.

WOOD, C.J.B. (1976), Conflict in resource management and the use of threat: The Goldstream controversy, *Natural Resources Journal*, 16(1): 137-158.

WOOD, H.A. (1956), The influence of the St. Lawrence Seaway upon the Cornwall-Prescott Area, Ontario, *The Canadian Geographer*, 7: 31-34.

5

ASPECTS OF RESIDENTIAL SUBDIVISION PLANNING FOR ENERGY CONSERVATION

Keith Nicol

Urban planners are responsible for designing and regulating development in the urban areas of Canada. Since the early 1970s when energy prices rapidly began to increase, planners have slowly begun to focus more attention on how to reduce overall urban energy consumption through more careful planning and design. The physical environment induces an energy demand which can be modified by planning urban form and land use to minimize demand. For example, concentric ring cities use the most energy, linear cities the least and polynucleated cities fall in between (Canada, 1982). The largest potential for planners attempting to conserve energy in urban areas is to reduce the use of energy in transportation and in buildings. Transportation consumes roughly 25% of Canada's total energy budget, and buildings about 35% (Land and Armour, 1982). Thus a variety of planning measures have been proposed which reduce the demand for energy. A compact contiguous urban form can create land use patterns that minimize energy used in transportation and might include policies which inhibit urban sprawl, concentrate development in nodes, create more local recreational opportunities, and develop local street patterns which minimize expenditure on transportation. Energy use by buildings may be reduced through planning energy-efficient neighborhoods and encouraging site planning in response to local climate. Solar orientation of streets, buildings, and landscaping, and design of energy efficient building forms which also protect solar access might be ways in which planners could reduce energy utilized in buildings, particularly in residential neighborhoods. This paper explores one aspect of this last series of planning measures--the effect of building orientation on energy use in twelve Canadian cities.

HISTORY OF CANADIAN SUBDIVISION PLANNING (1946-PRESENT)

Past trends in certain aspects of subdivision planning constitute a perspective for energy conscious urban planning. The literature on subdivision planning contains a limited amount of Canadian material. In the late 1940s and early 1950s one of the strongest influences on Canadian subdivision design was the city plan of Radburn, New Jersey (RAIC,1952). This subdivision's design (developed in 1928) demonstrated the concept of a whole neighborhood as a self-contained community unit and became the traditional way in which residential subdivisions were developed. The following principles were the basis of this plan:

1. diversified forms of housing accommodation;
2. the provision of schools, open space and some commercial activity within the project; and
3. the utilization of superblocks which separate main traffic from residential streets. Residential streets are predominantly composed of crescents, loops and cul-de-sacs.

Although these principles were not common in Canada in the late 1940s the early 1950s, certain developers and planners were utilizing similiar principles. The Wildwood housing development (started in 1946) in Winnipeg, Manitoba, utilized this approach except that loops were used instead of dead end streets because they suited Winnipeg's climate and the need for snow clearance (RAIC, 1952: 4). Prior to this approach, residential subdivisions in most Canadian cities were typically formed by grid street networks up to the 1950s or early 1960s. Orientation of subdivision streets was not considered to be significant in most cases, although it was suggested that row housing be constructed with an orientation of southeast to northwest or southwest to northeast (RAIC, 1952: 26). The general current recommendation for street orientation is east to west to ensure optimal energy conservation.

Curvilinear or Radburn design was popular enough by the late 1950s to warrent an article in *Ontario Planning* entitled "Subdivision Design - Good, Grid and Gimmicky." The "Good" plan exemplified the principles of the Radburn - design whereas the "Gimmicky" design was a confused layout based on "curves for the sake of curves" (Ontario Community Planning, 1958: 1-3). Because automobile collisions are most frequent at intersections, this type of street design reduces intersections and total road length to make a safer street pattern (UBC, 1965: 76). By the mid 1960s, safe street and intersection design were given a great deal of attention (Vogel, 1965: 44-45; UBC, 1965: 77). Lot pattern and street orientation were not considered very important (UBC, 1965: 77), at least in the Vancouver area:

> The lot design criteria which are significant are the frontage measured along the access street, the average width of the lot, the depth of the lot and the area of the lot.

By the late 1960s, the issue of urban sprawl became an important issue for some subdivision planners. By this time Radburn subdivisions were common in all parts of Canada and natural features like streams, groves of trees, and interesting topography were giving way to the bulldozer. Clustering or cluster development began to receive attention especially in the more populated regions of Canada. In North America, the Village Green cluster subdivision in Hillsborugh, New Jersey gained recognition with its enviable record of 40% of the total subdivision area devoted to public park and open space (Little, 1968: 45). Some Canadian planners in recent years have promoted clustering partially for the reasons described above but also because some buildings act as windbreaks for those nearby and reduce total energy demands (Hix, 1978: 42-44).

Another modification in subdivision design was introduced in the early and mid 1970s to combat the high cost of residential land. The Little River Acres subdivision in Windsor, Ontario was developed with lots measuring just 10 meters by 20 meters, roughly half the size of normal lots (Hickey, 1975: 24). To create reasonable amounts of open space and varieties of housing on small lots, the concept of zero lot line was introduced to permit each house to have one or more of its sides along the property line. Municipal zoning bylaws normally require houses to be set back prescribed distances from the property lines. However, with reduced lot sizes this places a large constraint on housing types and styles: "With smaller lots and the normal zoning setback requirements you start to get stereotyped little boxes--same shape, same size, same place on the lot" (Hickey, 1975: 26). Zero lot line zoning reduces this problem and also is a technique which municipalities are presently investigating as a way of reducing the shading of passive-or active-solar heated buildings induced by traditional zoning setbacks (Macleod,

1983).

With the advent of higher energy costs, planners at all levels of government began to investigate methods by which energy use might be reduced. By the late 1970s and early 1980s, Canadian municipalities had started to introduce various energy conservation measures. The first Canadian bylaw protecting solar access was passed in 1978 in Millet, Alberta (Lands Directorate, 1983: 7). Brampton, Ontario in 1979 introduced a bylaw requiring that building and street orientation favor passive solar collection for the proposed Alliance subdivision. Grid street patterns were encouraged with the majority of streets oriented east-west. Bylaw 139-79 (Land and Armour, 1982: 157-159) stipulated that buildings:

> be constructed so that the main wall facing north shall not vary from the east-west axis by more than 20° north of due east (or 20° south of due west) or by less than 12° north of due east (or 12° south of due west).

Other municipalities began to follow Brampton's lead of energy-conscious planning. Champlain Heights, a new subdivision in Vancouver, British Columbia, was built on south facing slopes utilizing a majority of east-west streets to ensure maximum solar access (Land and Armour, 1982: 103-106). The Peace River Regional Planning Commission (1982: 2) includes the following introduction to their residential subdivision booklet:

> This booklet attempts to follow two main themes. The first is energy conservation. In Northern Latitudes such as ours it is important to ensure that the maximum amount of passive solar energy is capable of being utilized by each and every household. We will examine design elements at the neighborhood level that can serve to increase passive solar gain. The second dominating theme is the utilization of natural site features. Having roads following existing contours and using natural drainage patterns to handle storm water flow are just two examples of natural site utilization.

The South March Subdivision in Kanata, Ontario is composed of superblocks, oriented east-west with an emphasis on compact, higher density housing and large amounts of open space (Land and Armour, 1982:

83-88). The Bedford Park development in Bedford, Nova Scotia, incorporates planning measures previously utilized for other purposes: cluster development, zero lot line zoning, and buildings oriented to maximize passive solar heating.

The preceding analysis of subdivision design demonstrates that one of the prime techniques of municipalities to reduce residential energy consumption is to control street alignment. Generally an east-west orientation for subdivision streets favors passive and active solar collection by minimizing shading and maximizing the area of south facing windows. Little research, however, has been done on the effect of orientation of buildings to determine the amount of energy savings that might occur due to an east-west street pattern instead of an alternative street alignment. This problem is explored in the remainder of this chapter through examples from various Canadian locations.

THE INFLUENCE OF BUILDING ORIENTATION ON ENERGY CONSUMPTION

Many factors enter into the problem of building orientation and energy use. For instance, window size and placement, thermal mass, the amount of insulation and air tightness influence how much solar energy can be gained and how little building heat can be lost. Because of duplicity in housing construction and variations in how the occupants use energy, actual comparisons of energy use by identical buildings orientated in different ways are difficult to accomplish. An alternative, modeling energy use via a computer program, permits a great deal of flexibility in building design and facilitates heat-loss calculations when different orientations are examined. A building heat loss model entitled, "HOTCAN," has been developed by the National Research Council, Division of Building of Research in Saskatoon, Saskatchewan (Dumont, Lux and Orr, 1982). This program, designed for use with microcomputers, aids builders and architects in quick calculations of buildings' energy use. A standard house was used for this study (Figure 1) and its thermal performance was investigated at twelve Canadian locations (Table 1). To determine the influence of orientation, the house was "rotated" through the four cardinal directions. A variety of "standards" was used to determine the effect of changes in levels of insulation and air tightness. The

TABLE 1

THE HEATING COST[1] ASSOCIATED WITH ORIENTATION OF BUILDING AND LEVELS OF INSULATION AT VARIOUS CANADIAN LOCATIONS ($)

Orientation	VANCOUVER	EDMONTON	SUFFIELD	SWIFT CURRENT	SASKATOON	WINNIPEG	TORONTO	OTTAWA	MONTREAL	FREDERICTON	HALIFAX	ST. JOHNS
1978 STANDARDS[3]												
South	677	1584	1094	1372	1493	1531	1025	1186	1100	1173	989	1249
North	774	1773	1283	1559	1701	1720	1129	1319	1233	1320	1096	1340
East/West[2]	720	1671	1190	1450	1600	1623	1069	1243	1159	1236	1033	1280
1983 SUPERINSULATION LEVELS[4] (Values given in parenthesis are for non-shuttered windows with consequently higher heating costs)												
South	99(139)	417(505)	234(298)	333(411)	380(468)	403(491)	221(280)	276(345)	244(308)	257(323)	196(249)	273(337)
North	156(202)	538(633)	338(410)	447(533)	509(603)	525(619)	297(363)	367(443)	326(398)	357(430)	262(321)	342(409)
East/West[2]	133(177)	491(583)	291(360)	397(480)	460(552)	474(566)	260(323)	327(400)	291(360)	314(384)	227(283)	306(372)

[1]Assumes house is heated by electricity costing 5¢ kilowatt/hr. (the average price for electricity, 1 January, 1983).

[2]Although east and west orientations do not always produce equal results, they generally differ by only a few dollars.

[3]The 1978 National Standards include the following insulation levels:

ceiling RSl - 5.4

walls RSl - 2.6

basement walls below grade RSl - 1.4

[4]The superinsulated house typical of some houses now being built in the prairie provinces has insulation levels:

ceiling RSl - 10.7

walls RSl - 7

basement walls below grade RSl - 2.5 as well as triple glazing and an air-to-air heat exchanger.

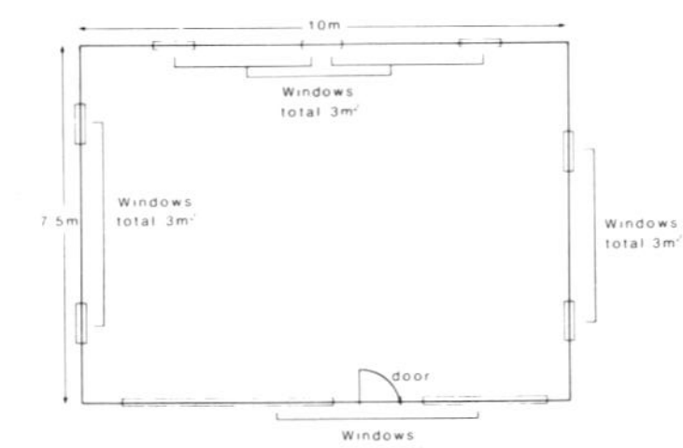

Figure 1. Illustration of Standard House used in "HOTCAN" Program to determine Heating Costs.

Main Building Components (main floor, basement) have total area of 150 m^2

Variations in Insulation used in Model.

Insulation (RSI):	1978	1982	Superinsulated
Walls	2.6	3.5	7
Ceiling	5.4	8.5	10.7
Basement Walls Below Grade	1.4	2.5	2.5
Windows	Double Glazed	Double Glazed on South; Triple Glazed on others	Triple Glazed

Note: No "north" arrow is indicated because house is "rotated" through orientations toward north, east, west, and south to determine the influence of orientation.

insulation levels ranged from the 1978 National Building Code Standards to what is considered to be good building practice today in the severe-weather Prairie region of Canada (Cole, 1983). The latter include such items as an air-to-air heat exchanger, window shutters, and triple glazed windows. Although such a building is unlikely to be cost effective at present in certain parts of Canada (e.g., Coastal British Columbia), uniformity was maintained to permit comparison among all places. An intermediate level of air tightness and insulation which is more typical of building standards across Canada today was also investigated (Cole, 1983). (Present recommended building standards, not included in Table 1, have heating values approximately $150 to $200 more than those for the superinsulated house.)

The effect of orientation varies significantly across Canada. Because of a higher heating load induced by a severe, but sunny winter climate, the Prairie provinces show the largest impact of orientation. According to 1978 standards, a north-facing house in Suffield, Alberta is estimated to pay 17% more for heating than that of the identical south-facing dwelling. In Saskatoon, this difference translates into an annual heating difference of $208. The effect of orientation is less influential where winters are less severe (e.g. Vancouver) or in over-cast environments where the heating season extends throughout most of the year. Thus although north-facing glass receives proportionally more energy (e.g., St. John's), and although heating costs are high, the difference between north and south orientation is calculated to be just 7% (using 1978 building measures). At locations between these climatic types (cold, sunny winters or extended heating seasons under overcast conditions), orientation produces intermediate energy savings. In Southern Ontario, Southern Quebec and the Maritimes, orientation (south-facing versus north-facing) produces calculated savings of 10-13%, equivalent to saving an annual cost of $104-147. Orienting the standard house east or west tended to result in almost identical energy costs. The maximum difference, in Swift Current, was a yearly total of $13. For the 1978 standards, houses not oriented to the south pay between 3 & 9% more. This amount aquals in actual energy costs from $30 to $110 more annually.

Heating was assumed to occur via electricity because it is the only heating method used or available country-wide. Natural gas is not available in the Maritimes, and heating oil is rarely used in Alberta. The electricity was assumed to be 100% efficient and to cost 5¢ per kilowatt-hour. Although Manitobans pay as little as 3.9¢ per kilowatt-hour and residents of Prince Edward Island pay 12.2¢ per kilowatt-hour, a population-weighted average using January 1983 data gives a mean cost of 5¢ per kilowatt-hour (Hillen, 1983). Heating by a less expensive technique (e.g., natural gas or wood) would reduce the impact of orientation, while heating with a more costly method would accentuate the effect of housing orientation. On the basis of 1978 building standards, solar energy is calculated to contribute anywhere from 27 to 35% of the required space heating of the standard house oriented toward the south (Figure 1). This share drops to a range of 18 to 26% if the same house is oriented toward the north. Solar energy contributes from 24 to 31% in a house oriented east or west depending on which of the twelve locations is examined.

As insulation standards are increased and air tightness of the buildings is improved, the effect of orientation is reduced. The lower part of Table 1 illustrates the pronounced effect of super-insulating Canadian homes on heating costs. This suggests that, as building standards improve, street orientation will have an increasingly smaller impact on energy use by buildings. Although all cases increased (e.g., for Vancouver, a north-facing home would use 57% more energy than an identical south-facing building), actual energy costs were smaller. This difference for Vancouver amounts to just $57 and indicates how, from a planning standpoint, given present energy costs, far more energy can be conserved by higher insulation levels or better building codes than through control of street patterns. With super insulation, including a night-time window shutter system, solar energy is calculated to satisfy anywhere from 49 to 66% of the annual heating requirements using the south-facing standard house. Even if this house faces north, the sun would likely supply 34 to 47% of the total heating needs depending on the building's location. Extra insulation, air-to-air heat exchanges, and triple glazing windows are costly and builders and developers will likely resist any regulations which increase the cost of their product. Furthermore construction standards are generally beyond the control of municipal or regional governments. Energy conscious planners, therefore, may have to settle for measures which, although not as effective as those they might wish to employ are more acceptable politically.

To assess the overall impact of street orientation on energy use, other factors also must be considered. The effect of future increases in the costs of energy will tend to *accentuate* the influence of orientation, in spite of the construction of better-insulated houses. With a doubling of electrical costs, even a north-facing house constructed to 1983 superinsulated standards would pay at least 25% more (generally 30 to 40% more) than its south-facing counterpart, costing up to $260 annually. Furthermore because houses and subdivisions are built with lifespans of at least 25 years, these savings are cumulative. And if street orientation is appropriate in the first place, then adding active solar collectors for space heating and hot water at some future date is much more economic than attempting to retrofit a poorly-designed subdivision with solar collectors.

CONCLUSIONS

Our communities in the past have been shaped and controlled by many factors. Energy was not traditionally one of these factors because it was relatively inexpensive. Consequently our settlements are better suited to a supply of cheap energy than to today's situation or to that predicted for the future. Local governments, however, through control of city plans and bylaws for subdivision and zoning can produce urban patterns which encourage energy efficiency. Controlling favorable street orientation is one increasingly common technique whereby Canadian municipalities reduce residential heating costs. A grid or semi-grid pattern of east-west streets produces the type of lot most suitable for passively solar heated buildings. This pattern minimizes shading and allows large amounts of south-facing glass to be used. This road lay out is in fact similar to residential street design prior to 1945 when the now popular pattern of cul-de-sacs and crescents (Radburn design) was introduced in Canadian cities. Planners are also beginning to make use of other familiar principles such as clustering and zero lot line zoning to reduce residential energy use.

On the basis of a standard house (Figure 1) and climatic information for twelve locations across Canada, the effect of building orientation on energy consumption was assessed by computer program. The results (Table 1) illustrating the impact of orientation and levels of insulation permit the following conclusions to be drawn.

1. The effect of orientation is quite variable for those cities tested; it reaches a maximum on the Prairies, is intermediate in eastern Canada, and is least pronounced in the coastal areas. Energy efficiency, but one aspect of subdivision planning, can not be effectively represented by a single number or percentage (Table 1).
2. The influence of upgraded standards of insulation and air tightness far outweigh (Table 1) the influence of orientation, and well-insulated houses reduce the effect of orientation. Building codes, however, do not normally lie within community control.
3. When energy prices increase, the influence of orientation is accentuated. Savings in energy cost are accumulated over the lifespan of the house and favorably oriented buildings likely can be more easily adapted to active solar collectors if changes in future production of energy enhance their cost effectiveness.

Over 70% of the urban buildings likely to be needed by the year 2000 are already in place. Through careful planning, the remaining 30% could be designed and constructed to maximize energy efficiency (Government of British Columbia).

REFERENCES

CANADA (1982), *Energy and Urban Planning* (Ottawa: Energy, Mines and Resources).

COLE, R. (1983), University of British Columbia School of Architecture (personal communication).

DUMONT, R., LUX, M., and ORR, H. (1982),*"HOTCAN": A Computer Programme for Estimating the Space Heating Requirement of Residences* (Saskatoon, Saskatchewan: National Research Council, Division of Building Research).

GOVERNMENT OF BRITISH COLUMBIA (1982), *Fuel for Thought* (Victoria: Energy, Mines and Petroleum Resources).

HICKEY, B. (1975), Zero lot line--A unique concept in subdivision design, *Living Places*, 11(2): 24-30.

HILLEN, E. (1983), Variations on a theme of price tags, Maclean's (11 April).

HIX, J. (1978), South March energy conserving community, *Canadian Architect*, 42-44.

LANDS DIRECTORATE (1983), *Land*, 4(1) (Ottawa: Environment Canada).

LANG, R. and ARMOUR, A. (1982), *Planning Land to Conserve Energy* (Ottawa: Environment Canada, Report No. 25).

LITTLE, C. (1968), *Challenge of the Land* (London: Pergamon Press): 42-47.

MACLEOD, B. (1983), Ministry of Energy, Government of Ontario (personal communication).

ONTARIO COMMUNITY PLANNING BRANCH (1958), Subdivision design--good, grid and gimmicky, *Ontario Planning*, 5(8): 1-3.

PEACE RIVER REGIONAL PLANNING COMMISSION (1982), *Urban Residential Subdivisions*.

ROYAL ARCHITECTURAL INSTITUTE OF CANADA (RAIC) (1952), The housing project, *Royal Architectural Institute of Canada Journal*, 29(9): 3-26.

UNIVERSITY OF BRITISH COLUMBIA (UBC) (1965), *Residential Land Subdivision - A Physical Evaluation* (Vancouver, Planning Staff, Project No. 2).

VOGEL, J. (1965), *Design of Subdivisions* (Seattle: University of Washington, Bureau of Government Research and Services).

6

FOUR DECADES OF GEOGRAPHICAL IMPACT BY CANADIAN SOCIAL HOUSING POLICIES

Ann McAfee

Between 1949-1983 the Canadian government has funded more than 390,000 non-market or social housing units. The conditions of funding, number and location of units, and type of households receiving assistance have changed over time. The geographical impact and social equity of housing assistance are evaluated in this paper from a national and local perspective.

FOUR DECADES OF HOUSING POLICIES

Social housing programs are redistributive initiatives funded by governments to assist low income households. Social, public, community, or non-market housing are terms used interchangeably to describe housing for people whose needs for adequate and affordable shelter cannot be met through the housing market. Non-market units are built by government, non-profit, or cooperative organizations. They are operated, and assisted by government subsidies, to ensure a continuing stock of affordable rental housing. Nine percent of Canadian housing is sheltered from inflationary pressures through non-market programs.

This paper presents both national and local perspectives on the impact of social housing programs. National observations rely upon unpublished material from Canada Mortgage and Housing Corporation (CMHC) (CMHC, 1982) and from the British Columbia Housing Management Commission (BCHMC) (Cogan and Darke, 1983). Local perspectives are derived from data gathered by the author during a case study of non-market housing in one city, Vancouver (McAfee: 1983), and from an analysis of cooperative housing in Vancouver (Darke: 1983).

Analysis of four decades of Canadian social housing policies suggests that:

1. Social housing programs by and large are well focused. Most assisted households require subsidies to ensure

affordability and need special design features and services often not available in market housing. Changes to programs during the last decade however, have resulted in less assistance being directed to those households most in need;

2. For those households receiving assistance, programs serve to redistribute income. Only a small proportion of lower income households, however, live in social housing or receive shelter allowances. Programs thus are inequitable in that households with similar incomes do not receive similar assistance; and

3. The location of social housing broadly reflects the distribution of population in Canada, although slightly more units are available in larger urban centers and in areas where provincial or local housing agencies have been aggressive in using social housing funds.

Divided into three sections, the paper first provides an overview of the evolution of Canadian social housing programs and emphasizes program characteristics and distributional consequences during two phases: 1949-1969 (federal-provincial public housing), and 1970-1983 (community non-profit and cooperative housing). A case study of social housing in Vancouver comprises the second section; socio-economic characteristics of assisted households are compared in terms of funding program, management, location, and design of housing. The third section makes concluding comments on the geographical impact and social equity of four decades of Canadian social housing programs.

CANADIAN SOCIAL HOUSING POLICIES, 1949-1983

Governments intervene in the housing market to assist modest and low income households obtain adequate and affordable housing. Government assistance may comprise supply-side incentives, such as grants or loans to non-profit agencies who in turn build or buy and renovate housing, or demand-side programs which assist households through shelter allowances, rent supplements, and income support programs. Canadian housing programs have favored the supply-side by focusing on building incentives.

Canadian housing programs can be divided into two phases. From 1949-1969 federal-provincial partnerships (sometimes with city participation) built and managed public housing for low income families, seniors, and the disabled. Between 1969-1974 public housing programs underwent extensive evaluation after which the type of funding, intended recipients, and sponsoring agencies changed. Senior governments in 1974 redirected funding to permit such community groups as cities, non-profit organizations, and cooperative societies to build and manage social housing for a mix of modest and low income households.

Federal-Provincial Public Housing, 1949-1969

The first Canadian social housing legislation was introduced in 1949. Under the public housing sections of the National Housing Act, cost-shared, federal-provincial programs funded publically-owned and provincially-managed housing for low-income households. Subsidies ensure that tenants should pay no more than 25-30 per cent of their income for rent.

Approximately 205,000 public housing units have been built since 1949. Most of the units are still occupied and are managed by provincial housing corporations. Ontario has the largest share of public housing (43%), followed by Quebec (22%), Alberta (7%), Manitoba (7%) and B.C. (6%) (Cogan and Darke, 1983). The distribution of public housing reflects the aggressive housing programs of the Ontario government which were supported by sufficient provincial monies to take advantage of the programs' cost-sharing provisions (Rose: 1980). To a lesser extent, units have been allocated to areas with traditionally low vacancy rates (Vancouver) or with entrepreneurs reluctant to build rental housing (Newfoundland, Saskatchewan). Public housing units have been directed toward three consumer groups--senior (48%), families (45%) and the disabled (7%).

Although the public housing program provides tenants with affordable shelter, a retrospective look at the program reveals problems in impact, equity, cost, and community acceptance. In their study of housing managed by provincial housing agencies, Cogan and Darke (1983) found that a low income household's access to public housing varied according to its location in Canada. Cogan and Darke compared the number of public housing units in each province with the number of households below provincially-specific low income "poverty" lines, those comprising annual incomes between $8,300 and $12,200. Households with incomes less than the poverty lines were defined as "high priority" for public housing assistance or experiencing "core housing needs" unlikely to be met in market housing. On the basis of Cogan-Darke's measure relating supply to need,Canada has on average about one public housing unit for every ten low income households. The ratio varies from a low of one unit to every 30 low income households in British Columbia to a figure of one to five in Newfoundland. Little or no assistance is available for low income households unable to live in public housing. Several provinces began in the mid-1970's to provide shelter allowances for households renting market units.

Community acceptance of public housing has been mixed. Projects for senior citizens are welcome, but large developments for low income families are not. Fears of declining property values, inferior qualities of design, strain on community services, adverse socio-economic mixes, and establishing precedents for building yet higher density housing in lower density areas, are among the objections expressed to public housing.

To reduce community resistance, public housing is often located in areas of low desirability. A study by the City of Vancouver in 1978 showed that 20% of public housing was not located within easy access of public transit or shops (McAfee, 1979). Only 50% of all projects were located within five blocks of park or recreational facilities. Forty percent were located next to, or within, audible distance of major traffic corridors or railtracks. Six percent were located near industrial activities.

Concern for the cost of public housing, community reaction to developments, and the possible social consequences of large concentrations of low income households led to a review in 1969 of Canadian social housing programs. Changes to program delivery, funding, and tenant mix were recommended.

Revision, 1969-1973, and New Community Housing Programs, 1974-1983

New directions in social housing emerged in the late 1960s in response to an increasing awareness of urban problems, concern for livability, and recognition of the continuing cost of providing deep subsidies. Between 1969 and 1973 a National Housing Conference, a Task Force on Housing (Canada, 1969), and an evaluation of social housing programs provided the basis for amendments to the National Housing Act (Wheeler, 1969; Canada, 1969; Dennis and Fish, 1972). Existing public housing, managed by the provinces was to continue providing accommodation for low income households. New social housing was to be built by community groups such as municipalities, non-profit organizations, and cooperatives. The legislation encouraging consumers to be involved in the design and management of housing, also required a mix of modest and low income households which would provide social diversity, mitigate community resistance to public housing, and reduce the cost of subsidies.

Funding for the new community housing programs is provided mainly by the federal government with some assistance from the provinces and cities. From 1974 to 1979, federal assistance included a 10% capital grant and mortgage financing at 8% for 35 years. Federal grants since 1979 have assisted with the reduction of mortgage interest rates on units to as low as 2%. After receiving mortgage subsidies, tenants must cover all remaining operating costs with their rent payments. As federal assistance is slowly withdrawn, residents who can afford to pay gradually assume responsibility for larger proportions of the full mortgage and operating costs. Eventually only the lowest income households receive subsidies.

More than 185,000 units were provided between 1974 and 1983 by non-profit organizations and cooperatives. These units are for families (50%), seniors (40%), and others (10%). When the amount of community housing is compared to the numbers of low income households by area, the relative accessibility of this housing varies from 7 units per low income household in British Columbia to 13 units per low income household in

Newfoundland.

A review of social housing built between 1979-1982 provides a profile of participants and accomplishments (CMHC, 1982). In the years 1980 and 1981, social housing represented 13% of all dwelling unit starts in Canada and 17% of all rental starts. The contribution of social housing to the national or local housing stock however, varies over time and space, and its relative importance depends largely upon the activity of the housing market. In Vancouver, for example, social housing represented 13% of completions in 1980, but 64% in 1974. In the period 1980 to 1981, social housing exceeded 40% of all starts in the depressed markets of Winnipeg, Hamilton, and Oshawa. In active markets, such as Calgary and Edmonton, social housing represented less than 10% of starts. CMHC defines core need as the number of renters unable to obtain adequate housing at less than 30% of income. Most areas of Canada receive social housing allocations in proportion to core need although, in the period 1979 to 1982, a slightly higher proportion of units went to British Columbia and Quebec than indicated by core need, while Ontario received slightly less than its share of core need. The allocations to Quebec were made in response to that province's active programs, whereas the share accruing to British Columbia reflected that province's low vacancy rates and active third sector (cooperatives and non-profit organizations).

Not all social housing units, however, are directed to those with the greatest need. Using CMHC's definition of core need, only 37% of social housing built between 1979-1982 assisted people with incomes in the lowest quintile. The proportion of low-income households served varies on a project-by-project basis and is best understood by considering the impact of social housing programs on a specific housing market.

NON-MARKET HOUSING IN VANCOUVER: A CASE STUDY

As a result of substantial public debate about who benefits from non-market housing programs, the City of Vancouver Planning Department studied 7,000 non-market households (discussed in McAfee, 1983). The study reviewed units assisted by the city and built through public housing and community housing programs. Ninety percent of non-market housing in Vancouver is built on city land and often provided at less than the market price. Of the 7,000 units, 3,850 house seniors, disabled, and other non-family households. All units are either managed by governments or are non-profit projects which receive grants to cover operating deficits.

Based upon a market basket analysis of the cost of basic housing, utilities, food, transportation, and clothing,city staff defined high priority households as families with incomes less than $20,000, couples with incomes less than $15,000, and single persons with incomes less than $10,000. Medium to high priority households were defined as families with incomes between $20,000 and $30,000 and adult (childless) households with incomes from $10,000 to $20,000.

Who lives in Non-Market Housing?

The City of Vancouver has assisted 7,000 households receive non-market housing. Of these, one-third are seniors, one-third are families, and one-third comprise other singles and couples. Ten percent of all households include a disabled person. Half of the units are managed by government (public or municipal non-profit housing), one-third by non-profit societies, and the remaining 20 percent by cooperatives. A very high percentage of non-market households have low-incomes and special design or service requirements.

An estimated 50 percent of Vancouver households have incomes less than $20,000, compared to 87 percent of the city's non-market households. Statistics Canada estimates that, in British Columbia, one-third of all single-person households have incomes less than $7,000 and one-third of family incomes are less than $25,000. Eighty-nine percent of assisted households in Vancouver have incomes below the Statistics Canada "poverty lines" for relevant household sizes. Eighty-five percent fit Vancouvers definition of "need" based on market basket analysis. The most revealing test of whether programs are providing housing to needy households is whether they could afford (at 30% of income) similar-sized housing in the same area or elsewhere in the city. Less than 3% of assisted households in Vancouver could afford housing in the same area, and possibly as many as 10% could afford housing elsewhere in the city. Not all landlords, however, rent to families and disabled persons.

Cooperative housing is an alternative to home ownership. No more than ten percent of cooperative households could afford a low-priced suburban home. At most 2% might have sufficient income to purchase a modest priced house in the city provided they also possess funds for downpayment. Cooperatives, however, offer many of the benefits of home ownership in areas of the city where home ownerhsip is not affordable to modest income households.

Almost all households in Vancouver non-family developments (seniors, disabled) have very low incomes: less than 5% have incomes over $10,000. Incomes display considerably more variation among households in family developments.

Who lives in Family Non-Market Housing?

Family developments are projects in which the majority of units have at least two bedrooms and are grade-related (i.e. have access from the unit directly outdoors or via private or shared stairs down two to three floors to grade level). Most, but not all, households are families with children. Cooperatives encourage a mix of households including families, singles, couples, and seniors. Ninety percent of households in family projects fit into the household type and low-income categories identified as "needy". The remaining 10% are classified as lower priority households, and comprise singles and couples with incomes exceeding 20,000, and families with incomes exceeding $30,000. Most lower-priority households live in cooperatives and often pay higher rents which provide "in-project subsidies for lower-income households. Many of the households with higher incomes are large families with more than one wage earner. When incomes are adjusted to take into account the higher costs of living for larger households, about 5% of all non-market households and about 9% of families in non-market housing have incomes exceeding

Statistics Canada's low-income poverty line for their household size.

Variations in the proportion of priority households in family projects are explained by differences in funding programs, management policies, sponsor, location, and livability (Table 1). The effects of different funding programs and management policies are evident; the highest percentages of priority households live in public housing, the only family housing in British Columbia receiving both federal and provincial subsidies. Such "deep" subsidies allow very low-income families to be housed. Cooperatives and regional (Greater Vancouver Regional Housing--GVHC) non-profit family housing are not eligible for provincial subsidies. After receiving federal assistance, tenants must pay rent which cover costs. Cooperatives and non-profit housing are funded under similar community housing programs.

Differences in household type and in the proportion of low-income households are explained by variables other than funding program such as philosophy, management practices, and design. Government-managed housing (public housing and GVHC non-profit) give priority to families with children. Cooperatives encourage a mix of households. Only 60% on average of households in cooperatives comprise families with children. Differences in household incomes of cooperative and GVHC non-profit residents are best explained by management practices. The GVHC adjusts rents annually and places a surcharge on higher-income households to encourage some turnover of units. When units are vacated, the GVHC then selects tenants based upon need. In contrast, funding programs allow cooperatives to manage their rent levels, waiting lists, and selection criteria. Cooperatives for the most part, encourage members to stay as their incomes rise, and do not regularly adjust rents to reflect market rates. Some cooperatives select new members based upon need but most invite the next household on the waiting list to join. This procedure does not ensure that lower-income households receive priority in allocation of vacant units. If selection criteria stressed filling vacant units with needy families then the number of low-income households in most cooperatives could be 10 to 20% higher.

TABLE 1

CITY OF VANCOUVER NON-MARKET FAMILY HOUSEHOLDS,[1] 1981

Funding Program	Public Housing	Community Non-Profit	Housing Coops	City of Vancouver[2]
Manager	Province	GVHC[3]	Self[4]	--
Income				
$ 0 - 10,000	64%	12%	10%	10%
0 - 20,000	93	65	47	27
0 - 30,000	100	96	80	47

SOURCE: McAfee (1983).

[1]Families with children only. Public Housing and GVHC Units are totally inhabited by families. Cooperatives house a mix of family and non-family households. If all households are considered, then 15% of cooperative residents have incomes less than $10,000 and 52% have incomes less than $20,000.

[2]1981 Census of all family households in the City of Vancouver.

[3]GVHC: Greater Vancouver Regional District Non-Profit Housing Corporation.

[4]Each cooperative manages its own affairs.

The resident profiles of cooperatives vary greatly among projects. The proportion of low-income households varies from 50% to 100%. Cooperatives with a high proportion of needy households usually have been recently occupied, have higher mobility rates, have members with a strong commitment to housing the poor, or have higher densities.

Length of tenure (or mobility) is closely ralated to a project's proportion of low-income households. The newer the project, or the higher the mobility in an older project, the higher the proportion of low-income households. Mobility rates (turnover) in older (pre-1979) cooperatives are 5% per year, regardless of location or management. The 5% mobility rate is similiar to that of owners (5.6% move annually) but half that of renters (12% move annually). By comparison, GVHC units have mobility rates two or three times that of cooperatives which resemble those of renters in market housing. With low mobility rates, cooperatives do not provide many opportunities to house additional low-income families, and over time some cooperatives tend to "highgrade" and reduce their proportions of low-income households. Although encouraging a stable community, this practice reduces the stock of social housing available to low-income households.

High-amenity locations do not explain variations in proportions of low-income households among projects (Figure 1). Availability of city land is the most important locational influence. The city has provided sites in two high-amenity waterfront areas (Jericho and False Creek) and in a 178 ha "planned" community (Champlain Heights), while other projects have been scattered among older, often less attractive, east-side neighborhoods.

Developments managed by governments, regardless of location, house 95-100% households with low-to-moderate incomes. In the high-amenity False Creek area, the proportion of high-priority households in cooperatives varies from 50 to 85%. Age of development and nature of management policies best explain differences among cooperative profiles: newer and higher density developments house the most low-income households. Perhaps correlation between family housing built at densities exceeding 100 units per acre, and large proportions of lower-income

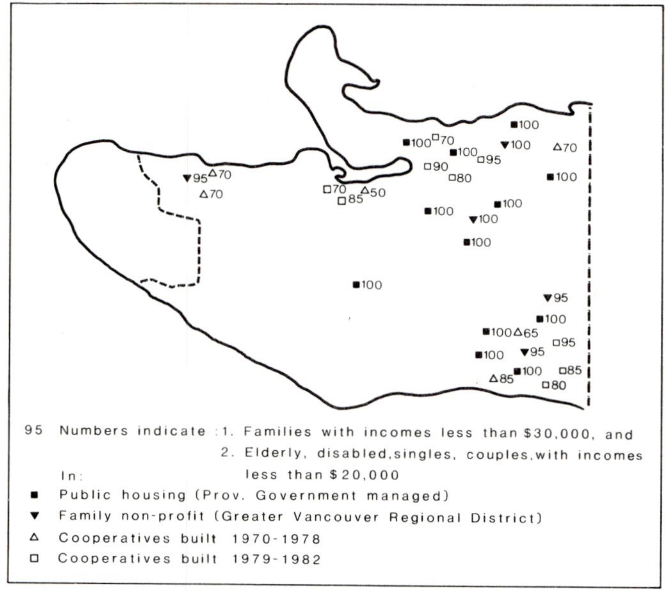

Figure 1. Location of Family Non-market Housing in the City of Vancouver, 1982

households may reflect qualities of livability, a subject which should be pursued in future studies.

The Vancouver case study demonstrates however, that, for the most part, social housing programs are well focused. Most units are occupied by households of modest means which, without assistance, would have a difficult time affording suitable market housing. Vancouver also displays however, negative characteristics which stem from the amount and conditions of funding, and philosophies of sponsoring agencies.

IMPACT AND EQUITY OF SOCIAL HOUSING PROGRAMS

During four decades, 390,000 social housing units have been built under the public housing (55%) and community housing (45%) provisions of the National Housing Act. Are the programs having the desired effect on those served and on the community at large, and are they reaching those in need? These questions can be addressed in terms of the programs' impact and equity.

Impact of Social Housing Programs

Impact may be assessed in terms of location of units, their effect on the housing market, and the extent to which those in need receive assistance. The location of social housing broadly reflects the population distribution of Canada; slightly more units are available in larger centers and in areas where provincial or local housing agencies have been aggressive in their use of housing funds. Within cities, units tend to be located on sites owned by local authorities, and in apartment areas where neighborhood outcry against social housing is not intense.

Social housing programs have provided residents with affordable housing, but have not reduced costs for households other than those in social housing units. If non-market housing comprised a sufficiently large proportion of the rental stock, then the cost of all housing could be reduced. Government subsidies would thus exert leverage beyond the number of units actually assisted. Such leverage has not occurred.

Non-market housing, about 9% of the Canadian housing stock, is insufficiently available to influence general prices. Only in specific markets (such as skid road inner-city areas) has non-market housing become a significant proportion of available stock.

Subsidies to non-market housing are high in the early years. Units protected from inflationary pressures, over time become more affordable than market housing, and ensure that a stock of housing is available to lower-income households. For programs to have a maximum impact, a high proportion of units should be occupied by lower-income households.

National studies and the Vancouver survey indicate that units are appropriately focused. Over 95% of social housing is occupied by low-income households. The proportion of low-income households varies, however, among different types of social housing. Public housing, seniors' non-profit accommodation, units for the disabled, and municipal non-profit dwellings are almost exclusively occupied by low-and-moderate income households.

The Vancouver study showed that not all cooperatives are appropriately inhabited. Philosophy and funding programs both promote an unsatisfactory income and household mix. Subsidies in early years are insufficient to allow community housing to accommodate a high proportion of low-income households and still recover operating costs. In later years, highgrading may result. These characteristics could be offset by modest changes to cooperative operating procedures which would increase their shares of low-income households.

The social benefits of income mix needs further analysis. Some evidence already suggests, however, that small (less than 25 units) projects of predominantly low-income households are sufficiently large to provide support services while still being small enough to integrate into existing communities (McAfee, 1978). At present, goals of income mix are achieved at the neighbourhood scale, but housing assistance is directed to those in need.

Social Equity of Non-Market Programs

Questions of social equity ask whether people with similar needs are being treated equitably. Equity can be evaluated within programs, and between recipients and non-recipients of programs' assistance.

Vertical and horizontal equity should be achieved within programs. Horizontal equity implies that households with similiar incomes receive similiar assistance. Vertical equity occurs when lower-income households receive more assistance than higher-income households. A comparison of rent-to-income expenditures between people in different housing programs indicates that households earning the same income do not necessarily face the same housing costs. If programs are horizontally equitable, households in the same circumstances, with the same incomes, would be assisted to the same extent. Although horizontal equity can never be totally achieved, variations in existing levels of assistance are substantial. Public Housing tenants pay no more than 25 to 30% of their income for housing. In Vancouver, 13% of cooperative and 37% of GVHC non-profit residents pay more than 30% of their income for housing; differences are explained by variations among subsidy programs.

Public housing, municipal housing, and post-1979 cooperative housing programs are vertically equitable in that the average monthly benefit per unit declines as income rises. In cooperatives built between 1974 and 1979, average benefit per unit remains fairly constant among classes of income.

Horizontal equity does not exist between people living in market and non-market housing. A study of low-income households in Vancouver found that 25% of households in market units have problems with affordability (McAfee, 1980). About 55% of single-person households and 86% of all renters with incomes less than $10,000, pay more than 30% of their income for housing. By comparison, residents in seniors and singles non-market housing pay no more than 25 to 30% of their income for shelter. Other 10% of families living in non-market housing pay more than 30% of their income for housing. By comparison, 25% of all families and 58% of single-parent families in market units have problems in affording housing.

CONCLUSIONS: FOUR DECADES OF SOCIAL HOUSING

During the past four decades social housing programs have provided funds to build non-market housing. Conditions of funding have changed from public housing to community housing yet the emphasis has remained on building or renovating units for lower-income households. Over 390,000 Canadian households have benefited from improvements in the quality and cost of their housing. Low-income households especially required assistance, but many households in similar circumstances still living in market units have received little or no assistance.

The desirability of continuing a supply-side approach to housing is undergoing review (CMHC, 1982). Governments are examining the ongoing cost of subsidies to non-market units. The inequitable treatment of market households with unrequited needs is also being examined. Projected declines in future demand for new housing may require less emphasis on supply-side housing programs. All these concerns may produce reductions in assistance to building programs but increases in demand-side subsidies such as shelter allowances which permit households to live in existing market units. Such changes will affect the type of units, dwelling location, assistance per household, and security of tenure previously established by social housing programs. Major modifications in existing programs could bring to a close, after four decades, a significant chapter in the provision of social housing in Canada.

REFERENCES

CANADA (1969), *Report of the Task Force on Housing and Urban Development* (Ottawa: Queen's Printer).

CANADA MORTGAGE AND HOUSING CORPORATION (CMHC) (1982), Section 56.1: Non profit and cooperative housing program evaluation, draft (Ottawa: CMHC).

COGAN, S. and DARKE, D. (1983), Comparative Canadian social housing characteristics, draft (Vancouver: British Columbia Housing Management Commission).

DARKE, D. (1983), Canadian federal housing cooperative program: Vancouver B.C., a case study, unpublished Master's thesis (Chapel Hill: University of North Carolina).

DENNIS, M. and FISH, S. (1972), *Programs in Search of A Policy* (Toronto: Hakkert).

MCAFEE, R.A. (1978), *Housing Families at High Density* (Vancouver: City Planning Department).

MCAFEE, R.A. (1979), *Understanding Vancouver's Housing* (Vancouver: City Planning Department).

MCAFEE, R.A. (1980), *A Social Housing Policy for Vancouver* (Vancouver: City Planning Department).

MCAFEE, R.A. (1983), *Who Lives in Non-Market Housing: An evaluation of the City of Vancouver Housing Program* (Vancouver: City Planning Department).

ROSE, A. (1980), *Canadian Housing Policies: 1935-1980* (Toronto: Butterworths).

WHEELER, M. (1969), *The Right to Housing* (Montreal: Harvest House).

7

STEMMING THE URBAN TIDE: SOME PROBLEMS OF LAND-USE PLANNING ON THE RURAL-URBAN FRINGE

Hugh J. Gayler

Since the Second World War, research in urban geography has become a major part of the disipline in Canadian universities. The spatial expansion of the city, resulting from the growth and decentralization of population and various economic activities has been one of the field's major concerns. A number of Canadian geographers have analyzed spatial variations in the nature and processes of urban expansion, including cities' impact on rural land uses, and dimensions of governmental policy and planning to reduce the conflict between urban and rural activities and to prevent the gradual alienation of rural land. This chapter examines contemporary problems of the rural-urban fring in Canada, especially the response to that problem in a small, yet important, agricultural area, the Niagara Region of Ontario, which is under constant threat from urban development.

RURAL-URBAN FRINGE STUDIES IN CANADA

Interest in various spatial aspects of urban growth and their conflicts with rural-based activities has not been the sole preserve of the urban geographer. Much of the early work came from geographers and planners whose major interest in resources and resource planning focused on agriculture, recreation and other rural land uses (Gertler, 1968; Irving, 1957; Krueger, 1959, 1965; Reeds, 1969). After the Second World War, increasing demand for housing accompanied by the decentralization of urban functions, growth in ex-urban living, and persistent weakness in the agricultural learning, encouraged a rapid and haphazard spread of low density urban development. A major transport revolution permitted an ever-increasing proportion of Canadian families to acquire a car, even a second car, and thus greatly improve their mobility; the political and planning climate unable and unwilling to prevent an onslaught on the countryside, encouraged urban expansion in the name of progress, homes for our heroes, growth, and increased wealth.

Expansion occurred on the margins of all major Canadian cities, but much of the initial work investigated problems of the rural-urban fringe in Southern Ontario and British Columbia. Until the 1970s, and rapid growth in Alberta, these two regions witnessed the fastest growth of population and urban-related activities in Canada, often at the expense of Canada's most valuable agricultural land. The loss of that land in several cases meant the loss of a particular agricultural economy, such as tender-fruit production and to a lesser extent, grape production, dependent on certain soil and climatic conditions which preclude the movement of these pursuits to other areas of the country.

Ontario's Niagara Region (Figure 1) has long received attention from geographers and planners. Not only interested in the processes influencing urban expansion and decline of rural-based activities, they applied their disciplines to develop an awareness of problems in government circles, and to formulate policies to help alleviate urban pressures on the countryside. Much of this attention focused on the so-called Niagara Fruit Belt, the area between the Niagara Escarpment and Lake Ontario, where haphazard urban growth along a major freeway threatened one of Canada's two tender-fruit growing areas. Although the early studies were undertaken on behalf of provincial government departments, however, their contemporary impact was negligible.

The very small amount of good agricultural land in British Columbia is located in the southwest, an area heavily influenced by the province's rapid post-war population growth and of particular concern to geographers and planners (Lower Mainland Regional Planning Board, 1956; 1963). Although the loss of land was serious and also adversely affected other agricultural areas, solutions to problems caused by urban expansion were not forthcoming by government; regional planning under an unsympathetic Social Credit Government was largely ineffectual.

The attention given to problems of the rural-urban fringe at universities in British Columbia, Ontario and elsewhere in Canada involved a growing number of graduate students; as many became teachers in other universities, especially in newly-established universities, or

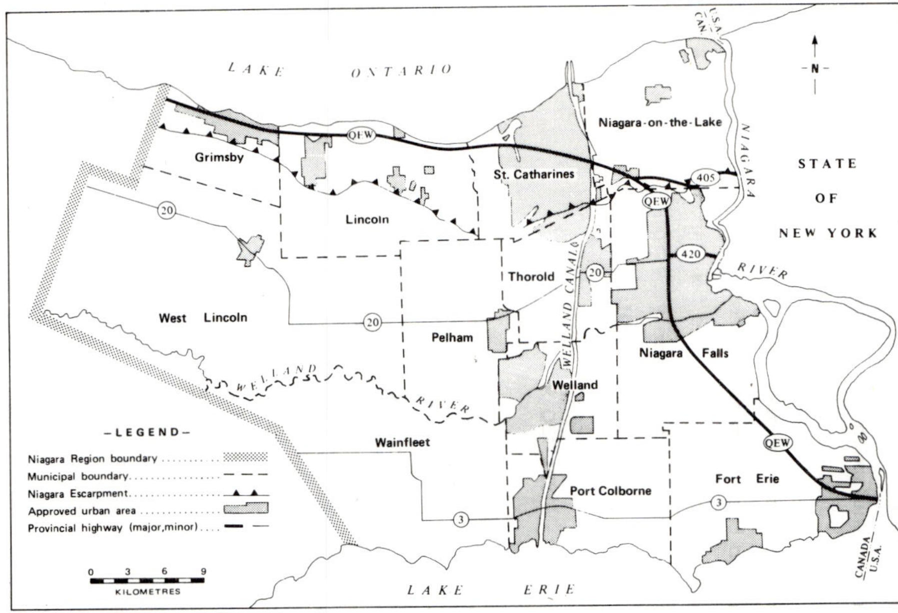

Figure 1. Urban Area Boundaries (UABs) of Ontario's Niagara Region, 1981. (Source: Regional Municipality of Niagara, 1981.)

went to an ever-increasing number of government positions, concern for the problems was reasonably expected to multiply. An extensive literature on the rural-urban fringe has been established since the late 1970s in Canada; many universities and government departments have developed specific programs both pure and applied to deal with problems of the rural-urban fringe and a strong lobby has slowly developed to persuade various levels of government to adopt stringent measures protecting the countryside from urban encroachment (Beesley and Russwurm, 1981; Furuseth and Pierce, 1982; Russwurm, 1977). Canadian geographers concerned with research, teaching and policy planning in this area established a specialty group in 1981 within the Canadian Association of Geographers.

PROBLEMS OF THE RURAL-URBAN FRING IN CANADA TODAY

The general message conveyed by most, if not all, geographers studying this problem is that, despite the many years of fairly intensive study by academics, involvement (and even sympathy) by government and substantial interest among members of the public in different parts of the country, the problems of urban expansion into the countryside continue to be acute and to necessitate persistent professional and public vigilance. Beesley and Russwurm (1981) show that the problems exist in all major urban areas of Canada, but only vary according to the rate of urban growth, the strength of the agricultural economy, and the effectiveness of planning authorities. The ubiquity of the problem reflects the fact that, most of Canada's urban areas are associated with the best agricultural land, and unlike some other industrialized countries, governmental land-use and policy planning has been either absent or ineffective in containing or directing urban development, strengthening the agricultural economy, and preventing conflict between urban and rural activities.

This absence or ineffectiveness of policy stems from a long tradition in North America whereby politicians have generally not interfered as processes of growth have affected the countryside (Gayler, 1979). In marked contrast, however, official plans and zoning by-laws in urban areas have extensively regulated land use. Areas away from the city have accommodated on a piecemeal basis various urban functions such as residential subdivisions, industrial parks, shopping centers, etc. These extend from the edge of the built-up area to distant low-density, sporadic, non-farm housing lots, severed from surrounding farms. These land uses comprise a variety of industrial, commercial and recreational functions designed for urbanites, but not contained within cities.

Urban and rural politicians, have never considered rural areas as sacrosanct for activities associated with agriculture. Agriculture frequently is not thought of as a permanent land use but is often classed as 'holding,' presumably until an urban use can be found. The readiness of rural areas to accept urban functions is seen as necessary for effective development by urban areas and any hindrances to growth are perceived as threats to urban standards of living. The politician's two strong allies, the developer and the farmer, have traditionally favored unrestricted growth in rural areas, especially during the period of rapid expansion in the 1950s and 1960s when both rapid windfall profits from the conversion of land.

Such rapid loss of agricultural land encouraged a number of moves to protect good land from urban encroachment. The arrival of an NDP Government in British Columbia in 1972, for example, led to the establishment of an Agricultural Land Reserve (ALR), a province-wide agricultural zoning covering 5 per cent of the land area and superseding all other land planning and zoning legislation (Furuseth and Pierce, 1982: 71-76). Despite the return of a Social Credit Government three years later, support for this agricultural zoning continues to be widespread and to generate public protest and publicity when developers attempt to persuade government to remove land from the ALR. The rate of loss of agricultural land in British Columbia undoubtedly has been considerably retarded since 1972.

In Ontario, on the other hand, the impetus for agricultural land preservation has emanated directly from the provincial government, but indirectly via the Planning Act and forms of approval for provincial municipalities' Official Plans. A major part of the planning especially in the urbanized areas of Southern Ontario, was facilitated by improvement in the structure of local government; at one time planning was hampered by many small town and township councils, and by larger cities administratively separate from their countries. The Ontario Government after 1969 established a number of two-tier regional and local municipal authorities, reduced the number of governmental units, and instituted machinery for comprehensive regional and local planning. Many geographers and professional planners set to work in this new climate, to protect the province's best agricultural land. Their efforts however, have been constantly thwarted by conservative local and provincial politicians intent on trying not to annoy unduly their traditional supporters in the agricultural and development sectors.

To enhance understanding of conflicts in the process of land management, the remainder of the chapter analyzes problems in the Niagara Region of Ontario. Through the lengthy actions of various interest groups, the Niagara Fruit Belt has become a cause célèbre (Gayler, 1982a). Despite some victories for those espousing agricultural preservation, the war between opposing groups is far from over, and various danger signals lurking on the horizon cause academics, planners, and the interested public to be constantly at the ready.

THE REGIONAL NIAGARA POLICY PLAN

The long and arduous fight to protect the best agricultural land and to restrict urban development can be seen in the protracted debate over the Regional Niagara Policy Plan. Although the various stages in formulation of this plan have been discussed fully elsewhere (Gayler, 1982b; Jackson, 1982), some aspects are recalled here because they influence strategies for future urban growth and preservation of agricultural land.

The Policy Plan, elevent years in the making (1970-81), was essentially completed by 1973, but was then delayed for almost eight years because agreement could not be reached with the provincial government as to where urban expansion would take place. Regional politicians, backed by their local counterparts, approved future urban areas that were grossly exaggerated in relation to likely future demands by the population, and that had little respect for good agricultural land. Their actions ensured that inefficient development, speculation, and uncertainty would continue to have harmful effects on agriculture. Politicians were unable, and often unwilling, to heed the advice of their own planners, provincial civil servants, and community public-interest groups. Serious conflict among the various parties culminated in a long and costly hearing by the Ontario Municipal Board (OMB) whose officers in effect determined the Urban Area Boundaries (UABs) shown in Figure 1. The UABs subsequently endorsed by the provincial government were not appealed by regional or local municipalities.

The Policy Plan, approved in 1981 and to be in place until 1996, must not be thought of, however, as the final word. The hearing officers at the time stressed the difficulties of establishing permanence (Ontario Municipal Board, 1979, 1981), and the Plan already has been challenged or amended on a number of occasions. These changes not only reflect new demands and the need for planning to be responsive to changing conditions, but also a continuing attempt to circumvent the intent of the Plan, to win battles that were formerly lost, and to promote possible development with little regard for the agricultural resource base.

FUTURE UNCERTAINTIES IN THE PLANNING PROCESS

The fact that the Policy Plan is not and will not be, a permanent document raises a number of issues in the years ahead which geographers and planners must examine. The uncertainty posed by these issues for the protection of a valuable agricultural resource will require the continued vigilance of all groups striving to preserve farmland.

Upturn in Growth

The present rate of urban expansion is retarded by the state of the Niagara regional economy. Unemployment is high (15-20%) and may not decline appreciably with improvement in the national economy because of certain structural, not cyclical, trends in manufacturing employment, especially the need to attract new employment in sectors of traditional economic growth (Gayler, 1982b). The average annual rate of population increase currently is less than 0.5 per cent, and the attraction of the Niagara Region as a retirement area is more than countered by the loss of young, well-educated people from the area.

The pressure on agricultural land by new residential subdivisions, industrial parks and shopping centers, therefore, has been alleviated in the short term. This situation could easily change, however. If more young families stay in the area, or migrate to it, for example, more single-family housing will be demanded on the urban fringe. A shift from heavy industry to light industry and services could encourage a move away from the environmentally less-sensitive areas south of the Niagara Escarpment to choice locations at feeeway intersections and service roads along the Niagara Fruit Belt. There is the very real fear that the strip development west of Grimsby along the Queen Elizabeth Way to Toronto will be repeated east of the town.

Disregard for the Policy Plan

The Policy Plan, on paper is designed to handle any upturn in growth to the satisfaction of farmland preservationists. Land allocations to the Niagara Fruit Belt municipalities for development were considered to be more than generous; their UABs were viewed as permanent and new growth was encouraged to go south (Regional Municipality of Niagara, 1981). In practice, however, the intent of the Plan can be easily circumvented. Unless their permanence is prescribed by law, UABs can be easily changed at the regional level by an official plan amendment; such changes proceed unless challenged by the province or by the general public and adjudicated by the OMB. The UAB boundaries and the strict policies regarding severances of farmland for development outside the boundaries have already been challenged. Local and regional politicians, traditionally pro-development, tend to approve these changes, usually against the advice of their planning staff. One type of challenge, motivated by the desire for more prestigious industrial land, could get worse. Although land approved for industry already exists, it is not always perceived as being in the right area--the Niagara Fruit Belt.

The Spread of Services

Figure 1 gives a misleading impression that present and future urban development is contained. Krueger (1978) has shown that development across the Fruit Belt is extensive and that only a small proportion of the area can now be considered rural. The dispersed, low-density, urban-orientated settlements have encouraged the spread of urban services, the persistent upgrading of which can only threaten the agricultural base still further.

Upgrading of infrastructure by various levels of government has unintentionally placed large areas of agricultural land in jeopardy. In order to improve local water supplies, trunk lines have been laid across rural areas; conflicts have then arisen between munipicalities on rights to tap into a passing water-line. Much rural development and the expansion of the UABs thus are restrained not through a lack of services but by an absence of political approval. If past performances are any guide to the future, then some skillful footwork must be undertaken to protect rural areas from further development.

Speculation in Rural Land

Current research suggests that the threat to agricultural land and the tender-fruit/grape economy is serious not only beyond the UABs but also beyond the trunk service lines (Krushelnicki and Bell, 1983). Their research placing the price per ha paid for agricultural land into various categories, found that almost 20 per cent of land transactions involving over two hectares generated over $27,100 per hectare (in 1981 dollars). Exceeding the ability of bona-fide farmers to pay, these prices represent investment in land for anticipated non-agricultural uses. A further thirteen per cent of transactions recorded prices between $20,000 and $27,100 per hectare for land with high agricultural productivity or lucrative anticipated non-agricultural uses.

The formulation of strict UABs was expected to remove the influence of speculation from the price of rural land. Given the short period in which to compile recent evidence on this phenomenon, experience in other areas suggests that speculation will nevertheless continue to be a factor in land prices. While the Plan lessens the risk of making a poor investment and the chances of a high profit inside the UAB, it also increases the risk and the prospect for profit should permission be obtained for development outside.

The Continued Need for Policing

While investigation using land-registry data is needed for accurate measurement, speculation is often directly evident by the form of land management. Once land is bought by purchasers not directly or indirectly involved in agriculture, the quality of agricultural-land management, albeit a difficult factor to assess, likely will decline. Almost anyone with money in Canadian society can speculate in land. Speculation, however, can be reduced, even stopped, and investment directed toward purely agricultural purposes, if permission to change the land use from agricultural to non-agricultural is refused. Seeking such permission under the Ontario Planning Act is highly visible, and can easily be contested by the general public should politicians side with those whose interests are inimical to agriculture. Constant policing is required, however, on the part of public-interest groups, such as PALS (Preservation of Agricultural Lands Society) in the Niagara Region because recommendations by planning staffs supporting preservation can be too easily overturned by their political masters.

Such policing is necessary to ensure that adverse decisions then go to the O.M.B. for appeal; only in this way can the risks in speculation outside the UABs be made so great that potential investors will seek a better return on their capital in enterprises not related to agricultural land. The OMB, being quasi-judical, regrettably can involve considerable legal and professional expenditures by advocates of land preservation, whose policing actions to be successful require large sums of money to match those expended on legal and professional services by developers. In the Niagara Region Policy Plan deliberations, for example, the O.M.B. hearing officers pointed out that PALS was the only group representing the wider public interest; municipalities tended to be pro-development and the Ontario Government was not represented (Ontario Municipal Board, 1981).

CONCLUSIONS

This discussion shows that, although issues of land-use planning on the rural-urban fringe today are an important focus of Canadian geography, our awareness originated with the immediate post-World War II generation of geographers who saw especially in Ontario and British Columbia the loss of valuable agricultural resources through unplanned urban sprawl. Through their actions and those of their students, governments have been challenged to play a more positive role in the preservation of agricultural land. The Niagara Region demonstrates, however, that sufficient uncertainty still exists to keep academics and public-interest groups vigilent in their determination to prevent the despoilation of major agricultural environments.

REFERENCES

BEESLEY, K.B. and RUSSWURM, L.H. (editors) (1981), *The Rural-Urban Fringe: Canadian Perspectives* (Downsview, Ontario: York University, Atkinson College, Department of Geography).

FURUSETH, O.J. and PIERCE, J.T. (1982), *Agricultural Land in an Urban Society* (Washington, D.C.: Association of American Geographers).

GAYLER, H.J. (1979), Political attitudes and urban expansion in the Niagara region, *Contact (Journal of Urban and Environmental Affairs)*, 11: 43-60.

GAYLER, (1982a), Conservation and development in urban growth: The preservation of agricultural land in the rural-urban fringe of Ontario, *Town Planning Review*, 53: 321-41.

Gayler, (1982b), The problems of adjusting to slow growth in the Niagara region of Ontario; *The Canadian Geographer*, 26: 165-71.

GERTLER, L.O. (1968), *The Niagara Escarpment Study: Fruit Belt Report* (Toronto: Ontario Department of Treasury and Economics).

IRVING, R.M. (editor), (1957), *Factors Affecting Land-Use in a Selected Area in Southern Ontario: A Land Use and Geographic Survey of Louth Township in Lincoln County* (Toronto: Ontario Department of Agriculture).

JACKSON, J.N. (1982), The Niagara fruit belt: The Ontario Municipal Board decision of 1981, *The Canadian Geographer*, 26: 171-76.

KRUEGER, R.R. (1959), Changing Land-use patterns in the Niagara fruit belt, *Transactions of the Royal Canadian Institute*, 32: 39-140.

KRUEGER, R.R. (1965), The geography of the orchard industry of Canada, *Geographical Bulletin*, 7: 28-71.

KRUEGER, R.R. (1978), Urbanization of the Niagara fruit belt, *The Canadian Geographer*, 22: 179-94.

KRUSHELNICKI, B. and BELL, S. (1983), Identifying urban price shadow effects in the urban countryside of Niagara region, paper presented to the Canadian Regional Science Association Annual Meeting, Vancouver.

LOWER MAINLAND REGIONAL PLANNING BOARD (1956), *Urban Sprawl* (New Westminster: Lower Mainland Regional Planning Board).

LOWER MAINLAND REGIONAL PLANNING BOARD (1963), *Land for Living* (New Westminister: Lower Mainland Regional Planning Board).

ONTARIO MUNICIPAL BOARD (1979), Re: Niagara Planning Area Official Plan partial referral (Niagara Falls, Port Colborne and Thorold Urban Area Boundaries), *O.M.B. Report*, 9: 286-307.

ONTARIO MUNICIPAL BOARD (1981), Re: Niagara Planning Area Official Plan, *O.M.B. Report*, 11: 353-466.

REEDS, L. (1969), *Niagara Region Agricultural Research Report* (Toronto: Ontario Department of Treasury and Economics).

REGIONAL MUNICIPALITY OF NIAGARA (1981), *Regional Niagara Policy Plan*, Office Consolidation, April 1981.

RUSSWURM, L.H. (1977), *The Surrounding of our Cities* (Ottawa: Community Planning Press).

8

MAPPING METROPOLITAN CANADA

James W. Young

The obstacles to unity posed by the physical distribution of Canada's population have been for more than a century a major concern of national policies. The ecumene is a strip some 6500 km. long, and often only 200 km. wide. Canada, nevertheless, has overcome many of the problems posed by this near unidimensionality and forged an integrative national system. Factorial studies of the cultural and socio-economic attributes of places during the past two decades have demonstrated the multidimensional reality of Canadian society, and pointed particularly to an urban size or hierarchial dimension imposed on the original east-west physical structure and French-English dichotomy (King, 1966; Ray, 1971; Simmons, 1978).

Visual portrayal has considerable heuristic value for understanding these integrative structures. Nearly all published maps, however, portray physical space and emphasize the near-unidimensionality of the physical ecumene; tree-graphs of urban Canada tend to isolate only dominance. This chapter uses techniques of multidimensional scaling (MDS) to explore whether or not satisfactory visual images can be obtained from selected interaction data for metropolitan Canada to ensure objective, accurate representation of time space and of social space.

VISUAL REPRESENTATION OF CANADA

The efficacy of the visual image in representing abstract concepts is well-known. All geography instructors employ blackboard graphics and maps, even though the processes of perception and thinking are not clearly understood (Robinson and Petchenik, 1976: 68-85). Visual image does not substitute for all words; words are still required to guide viewers to the right set of propositions (Mills, 1981: 18). The thematic map, halfway between the pictorial and the symbolic, preserves physical reality while conveying abstract concepts.

The cartogram which consciously distorts physical reality, like the cartoon, attempts to reduce the "noise" of that reality and ease decoding of the abstract concept (Mills, 1981: 78-9). Kadmon (1982: 3) suggests that the most effective cartograms are those which preserve some isomorphic correspondence with physical space. Empirical work on representation of areal units supports this idea, and also provides experimental evidence of the superiority of the cartogram over the conventional thematic map (Griffin, 1983).

Despite the cartogram's efficacy, surprisingly few examples can be found of its use in mapping attributes of Canada. Examination of various editions of the *Atlas of Canada* and of provincial atlases published in the last quarter-century, shows that the number of cartograms could be counted on one hand. The isodemographic map of Canada (Skoda and Robertson, 1972) is one of the few well known cartograms. Although treegraphs of urban structure are common, they are designed to indicate only dominance although they often exhibit some isomorphic correspondence along the abscissa with physical space because of the strong east-west regionalization of the urban hierarchy.

Factor analyses have been applied to a variety of socio-economic scalar attributes of urban centers, regions, and census divisions (King, 1966; Ray,1971; Simmons,1978). Visual representations comprising thematic mappings of one dimension and graphs of two dimensions are frequently presented. The purpose of factor analysis, however, is to identify an interpretable set of dimensions, whereas "the goal of multidimentional scaling is to find a compact, accurate, geometric representation of the original 'distances' between items"(Napior, 1978: 168). Useful reviews of the technique and its uses are available (Golledge and Rushton, 1972; Kruskal and Wish, 1978).

MDS permits the researcher to specify the number of dimensions, and facilitates, by its measures of goodness of fit, evaluation of the minimum number of dimensions required. A non-metric algorithm, MINISSA, eases the computational task in this study by reproducing only the ordinal scaling of the original interactions. The need is obviated to specify functional relationship between interactions and listed distances. The result in many cases, comprises an output configuration which approximates the metric space of the original interactions. This chapter seeks an accurate representation in three or fewer dimensions of Euclidean space based on distances which are linear functions of the input data. The educational efficacy of the configuration is enhanced by simple shapes and patterns, permitting a viewer to assume linearity in the evaluation of relative distances between points.

TIME SPACE

The empirical work on time space and social space is confined to the 24 Canadian 1981 census metropolitan areas, a restriction necessitated by the availability of 1976-81 migration data used as a surrogate for social space. These metropolitan centers account for 56 per cent of the 1981 population, and are foci of 76 per cent of the Canadian population found in the urban-centred regions defined by Simmons (1977).

The fastest travel time in the summer of 1983 by common carrier and/or automobile between the city centers was estimated to the nearest quarter-hour. Estimates for access and waiting time at airport terminals were partly based on a study by the Canadian Transport Commission (1970). Travel between some metropolitan centers within central Canada is faster by automobile than be scheduled airliner. Variations in travel time according to direction of the journey are insignificant in virtually all cases. The effect of errors in estimating access and waiting times was evaluated by calculating MDS outputs for airline travel time and for total travel time. Differences between the two configurations are trivial, mainly because non-metric MDS generates robust configurations.

The two-dimenstional configuration of time space, and the displacement vectors of each center for its location in physical space, are shown in Figure 1. Ewing (1974) argued that objectivity in the transformation from physical space to time space is ensured by equating mean physical and time distances, superimposing their centroids, and then rotating one map on the other until the sum of the displacement vectors is minimized. Figure 1 equates physical and time distances

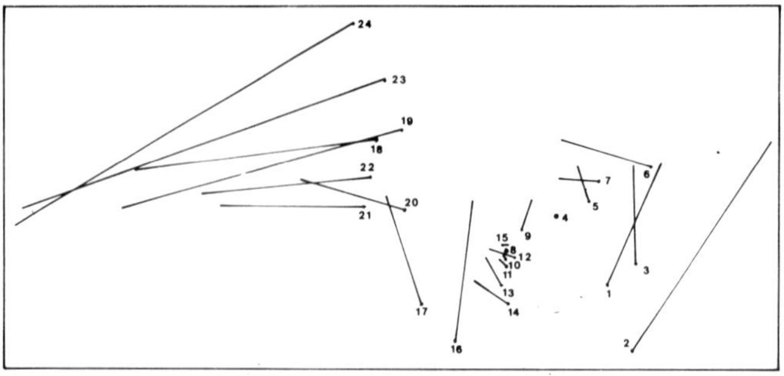

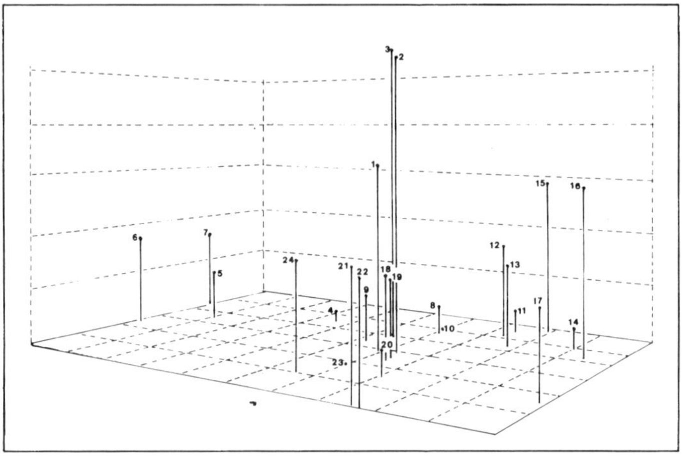

Figure 1. Time space of Metropolitan Canada.

1. Halifax
2. St. John's
3. Saint John
4. Montreal
5. Quebec
6. Chicoutimi-Jonquiere
7. Trois-Rivieres
8. Toronto
9. Ottawa-Hull
10. Hamilton
11. St. Catharines-Niagara
12. Kitchener
13. London
14. Windsor
15. Oshawa
16. Sudbury
17. Thunder Bay
18. Edmonton
19. Calgary
20. Winnipeg
21. Regina
22. Saskatoon
23. Vancouver
24. Victoria

Figure 2. Social Space of Metropolitan Canada. (Numbers correspond to those of Figure 1.)

between Toronto and Montreal, and superimposes the time space map on these twin foci to fix the rotation. These centers are the foci of travel within Canada; the Toronto-Montreal distance is near the point where surface and air travel times are equal. The displacement vectors consequently come close to representing the influence of modern air service in bringing about spatial convergence.

To fit a configuration to a "distance" matrix within a given dimensional constraint, the iterative procedure sought to minimize the Guttman-Lingoes coefficient of alienation (G-L coefficient) under semi-monotonic conditions (Golledge and Rushton, 1972: 23). Unlike weak monotonicity, semi-strong monotonicity does not permit the tied distances to exceed the number of ties in the input matrix. Hardly any difference was found in the outputs between the study's two monotonic options. Likewise, the differences between the goodness of fit measure (G-L coefficient) and the alternative stress measure devised by Kruskal were minor (Golledge and Rushton, 1972: 11).

The G-L coefficients for mapping onto 1, 2, 3 and 4 dimensions were 0.23, 0.13, 0.09, and 0.06 respectively. Kruskal's evaluation of stress indicates that the two dimensional fit is only "fair;" a "good" fit is not achieved in less than four dimensions (Golledge and Rushton, 1972: 12). To accept a configuration in m dimensions, evaluation should be made of the trade-off between parsimony, interpretability and visibility on the one hand, and goodness of fit on the other (Kruskal and Wish, 1978: 56). The choice of the two dimensional configuration is supported by a slight "elbow" exhibited in a plot of coefficients on dimensionality. A scatter diagram of distances and the original travel times confirms that linearity was preserved.

Time space was expected to be a U shape with the directly-connected major metropolitan centers delimiting the inner side. Transport innovations have resulted in greater time gains on the long haul, and these innovations are being applied to a physical layout which is almost unidimensional. Expectations were confirmed in Western Canada where the major centers are on the inner side, and the less well-connected centers (Regina and Saskatoon) are positioned by a relative outward push.

The eastern half of the U shape is formed, however, by the Quebec centers, with Atlantic Canada lying in the outer fringes. A replication of the Western pattern is not possible because of the need to accommodate poorly connected centers in Quebec along with the more distant centers of Atlantic Canada. The poor airline connections of Thunder Bay and Sudbury have resulted in a marked outward displacement to the fringe of the configuration. The similar displacement of Windsor occurs only when a two-dimensional physical space configuration is used as input. A random input configuration results in Windsor being inside the U shape, and having the same overall degree of fit. Tobler (1982: 8) argues that the problem of differing outcome configurations is avoided with input of a likely configuration such as that represented by physical space.

In view of the "fair" fit, correlations of travel time and derived distances were calculated for each ith center to all j centers: $j = 1, n; i \neq j$. The overall correlation for the 276 dyads is r = 0.95. The n center correlations as expected showed that the errors are generally distributed, except in the case of Windsor and Sudbury where the correlation coefficients are below r = 0.90. These centres are difficult to position in any generalized time space configuration because their airline connections are erratic, and produce only a weak relationship between travel time and physical distance. For the total of 276 dyads, the elasticity of travel time on physical distance is 0.44 (with r = 0.91) and reflects a greater long-haul time space convergence.

SOCIAL SPACE

Data on telephone calls, which have often been employed as a surrogate of social interaction, are now confidential. Use of migration data as a surrogate measure assumes that a migrant is usually necessary for the development of a social network of relatives and friends embracing two cities. McGahan (1982: 239), in summarizing the apparent scanty literature on inter-urban kinship patterns, points to the strong retention of these ties though, naturally, distance changes the forms and frequency of contact. Migration data do not allow for variations in either intensity and frequency of contact or degree of interest in the affairs of another center arising from an individual's social network. On the basis of some very limited telephone data of the late 1960s in Central Canada (Simmons 1972: 214), an association was sought with migration of that time. After removal of extreme residuals, the Pearson correlation coefficient rose to 0.84. Migration underpredicted telephone messages between close centers such as Kitchener and Guelph, Toronto and Hamilton. The social network of friends arising from the workplace and other organizations could easily extend over these distances without necessitating migration.

Any asymmetry in the size of the migrant's social network in the origin and destination centers was judged as unimportant. Consequently, gross flows between centers were used as the surrogate measure. Likewise, time decay effects in maintaining friendships over long distances were ignored. Social interaction is viewed here as an indication of the degree of interest and emotional ties with events in other places. Such an assertion requires further research, but if it is accepted as a working hypothesis, time decay affects on these phenomena could be minor as long as friends and kinfolk reside in the other center.

One approach in this situation is offered by Goodchild and Kwan (1978). Using data for news flow between 17 major Canadian centers, they extracted a dominance scaling and then applied MDS to the remainders of the interaction. Their result was a three-dimensional graph with dominance scaled on the vertical axis, and distance on the plane measuring spatial effects or impedance to interaction. As a visual representation of relationships, the graph loses heuristic appeal because it requires analytic calculations on two scalings in order to evaluate connectivity. Furthmore, the immediacy of the configuration's shape is lost in conveying an overall assessment.

The asymmetry problem in this study was evaluated by forming two impact matrices comprising impact on the larger and on the smaller center of each dyad, and then by calculating MDS configurations. In view of the robust nature of the configurations, they were not expected to differ significantly. The potential for differences is also reduced: whichever procedure is selected, the set of measurements for all but the largest and smallest centers is a combination of the ith center's impact on others, and their impacts on i. The mix depends on the rank size of the ith center.

The G-L coefficients for mapping impacts on the larger center into 1, 2, 3 and 4 dimensional spaces were 0.34, 0.25, 0.13 and 0.09 respectively. The coefficients for impacts on the smaller centers followed the same progression, but were between 0.03 and 0.55 higher. Both series exhibited an "elbow" in the progression at three dimensions. Figure 2 is a perspective graphing of the three-dimensional configuration of impacts on the larger center of each dyad. Whichever impact procedure is selected, the three-dimensional configurations are similar. Distances on the two configurations are linearly related with a correlation of $r = 0.89$. For each of the n centers, the correlations of the two sets of derived distances to j centers ($j = 1, n; 1 \neq j$) only fall below $r = 0.80$ in the case of the three Atlantic centers.

The fit of Figure 2 to the impact data can be described only as fair. The derived distances furthermore are not linearly related to the impacts. The best fitting function of distances to impact is logarithmic with $r = -0.88$. Figure 2 consequently exaggerates the small impacts (larger distances) relative to the larger impacts (small distances).

The social space of metropolitan Canada has a more complex configuration than that of time space. The centers of English Canada outline a rough egg-shaped space within which regional clusters, particularly Western Canada, are evident. The most visually striking element of the configuration is the isolation of Quebec, with Montreal acting as a bridge between the two solitudes of English and French Canada. Within English Canada space, size as well as regional effects are apparent, with the larger metropolitan centers drawn towards the cluster of those in Southern Ontario. This coalescing of larger centers is expected because, ceteris paribus, interaction is a multiplicative function of size of the origin and destination centers modified by spatial impediments (Isard 1960: 495-5). With distances from Toronto isolated, the five closest centers are all in Southern Ontario, but thereafter the ordinal sequence has the Alberta centers followed by Winnipeg and Montreal.

Metropolitanization of the population is an integrative process. Positioning relative to Toronto reflects impact of these centers on Toronto along with the degree to which their relationships to other centers replicate the pattern for Toronto. Oshawa for this reason, is pushed to the periphery because of its relative isolation from many of the metropolitan centers of Canada.

CONCLUSIONS

The objective mapping of time space is more successful than that of social space. Time space can be mapped into fewer dimensions, has a simpler shape and pattern, and distances are linearly related to time. From a heuristic viewpoint, the most severe problem of the social space configuration is its lack of linearity between distances and social impacts. This could be corrected by employing a metric MDS algorithm, but the errors in a three-dimensional configuration likely would increase.

Enlarging the sample to include additional smaller urban centers would increase the error of the output configuration within a given dimensional constraint, but not to the point where a visual image is necessarily discarded. The additional centers in time space are linked to the main network via connections through a neighboring major metropolitan center. A two-stage MDS procedure, in which the configuration of the major centers is calculated and then added as a fixed configuration into further interations to position the smaller centers, ensures that the latter do not determine the global configuration. Smaller centers should form clusters focused on the appropriate connecting centers which delimit the inner side of a U shaped zone. Errors would appear as distances between two neighboring clusters reflect the travel time between their metropolitan coci. Results in the case of social space are less predictable. Additional small centers with similiar patterns of little impact on those of Figure 2 might form a cluster located in the periphery. Such a cluster, however, would be misleading insofar as intracluster distances would suggest strong mutual connections between its members.

A more pressing line of research than further objective mappings, however, is to test empirically the heuristic value of cartograms and graphs. With criteria of "understanding" defined and operationalized, mappings similiar to those of this study could be tested; alternative forms of presentation could combine maps of physical space with matrices and written evaluations of interactions. The popularity of physical-space maps, however, probably reinforces a view favoring province-building at the expense of nation-building.

ACKNOWLEDGEMENTS

Thanks are extended to Concordia University for supplying computer time, and to the Demographic Section of Statistics Canada for estimates of 1976-81 migration. Dr. G.O. Ewing of McGill University provided a copy of the MINISSA program.

REFERENCES

CANADIAN TRANSPORT COMMISSION, RESEARCH BRANCH, (1970), *Inter-city Passenger Transport Study* (Ottawa: Queen's Printer).

EWING, G.O. (1974), Multidimensional scaling and time-space maps, *The Canadian Geographer*, 18: 171-7.

GOLLEDGE, R.G. and RUSHTON, G. (1972), *Multidimensional Scaling: Review and Geographical Applications* (Washington, D.C.: Association of American Geographers, Commission on College Geography, Technical Paper No. 10).

GOODCHILD, M.F. and KWAN, M.Y.C. (1978), Models of hierarchially dominated spatial interaction, *Environment and Planning A*, 10: 1307-17.

GRIFFIN, T.L.C. (1983), Recognition of areal units on topological cartograms, *The American Cartographer*, 10: 17-29.

ISARD, W. (1960), *Methods of Regional Analysis: An Introduction to Regional Science* (Cambridge: The M.I.T. Press, Regional Science Studies Series No. 4).

KADMON, N. (1982), Cartograms and topology, *Cartographica*, 19(3 and 4): 1-17.

KING, L.J. (1966), Cross-sectional analysis of Canadian urban dimensions: 1951 and 1961, *The Canadian Geographer*, 10: 205-22.

KRUSKAL, J.R. and WISH, M. (1978), *Multidimensional Scaling* (Beverly Hills: Sage Publications, Sage University Paper Series No. 11).

McGAHAN, P. (1982), *Urban Sociology in Canada* (Toronto: Butterworth and Company).

MILLS, M.I. (1981), *A Study of the Human Response to Pictorial Representation on Telidon* (Ottawa: Department of Communications, Telidon Behavioural Research 3).

NAPIOR, D. (1972), Nonmetric multidimensional techniques for summated ratings, in R.N. Shepard, A.K. Romney, and S.B. Nerlove (editors), *Multidimensional Scaling: Theory and Application in the Behavioral Sciences* (New York: Seminar Press): 157-78.

RAY, D.M. (1971), *Dimensions of Canadian Regionalism* (Ottawa: Information Canada, Geographical Paper No. 49).

ROBINSON, A.H. and PETCHENIK, B.B. (1976), *The Nature of Maps* (Chicago: University of Chicago Press).

SIMMONS, J.W. (1972), Interaction among the cities of Ontario and Quebec, in L.S. Bourne and R.D. MacKinnon (editors), *Urban Systems Development in Central Canada: Selected Papers* (Toronto: University of Toronto Press): 198-219.

SIMMONS, J.W. (1977), *Migration and the Canadian Urban System: Part I, Spatial Patterns* (Toronto: University of Toronto, Centre for Urban and Community Studies, Research Paper No. 85).

SIMMONS, J.W. (1978), *The Great Canadian Factor Analysis. Dimensions of Variation in the Canadian Urban System* (Toronto: University of Toronto, Centre for Urban and Community Studies, Research Paper No. 93).

SKODA, L. and ROBERTSON, J.C. (1972), *Isodemographic Map of Canada* (Ottawa: Information Canada, Geographical Paper No. 50).

TOBLER, W.R. (1982), Surveying multidimensional measurement, in R.G. Golledge and J.N. Rayner (editors), *Proximity and Preference: Problems in the Multidimensional Analysis of Large Data Sets* (Minneapolis: University of Minnesota Press): 3-9.

9

"WHERE IS HERE?" THE USES OF GEOGRAPHY IN ENGLISH-CANADIAN LITERATURE[1]

Edward M. Gibson

Those familiar with Canadian literature will recognize the "Where is here" in the title of this chapter as a line quoted from the critic Northrop Frye's (1971b), *The Bush Garden. Essays on the Canadian Imagination.* Northrop Frye, a trained literary critic, was not leaving his field of inquiry to engage in a question of location normally handled by geographers. Similarly, my purpose as a geographer is not to apply Northrop Frye's craft of literary criticism. After studying Canadian geography under Lewis Robinson and participating in this volume honoring his highly visible contribution to the writing of Canadian geography, my turning from geography to the untested craft of literary criticism may seem strange indeed. Here, however, I want to pursue a line of thought that has grown from my connection with Lewis Robinson since 1962, and that forms a bridge from Canadian literature to Canadian geography.

IMAGINATION AND GEOGRAPHY

Knowing Lewis Robinson's introductory lectures on the regional geography of Canada, and being familiar with his writings, you will recognize, his special interest in Canadian imagination. You will also see the starting point of my interest in both imagination and geography. Lewis Robinson says in effect at the beginning of his course, "On a blank piece of paper and without any reference to other material draw an outline map of Canada showing the provinces and the cities you feel are of importance to you." In other words, "What geography of Canada can you imagine?" Such a question is close to my preoccupation in teaching and writing in both fine arts and geography since 1962.

The life of Lewis Robinson is closely related to the purpose of this chapter. By examining his writing community, his peers, and his family, you see a tributary flowing into the mainstream of Canadian literature, our imaginations recorded in fiction, poetry, and plays. Talonbooks can be said to have grown from 1963, the year his son David and a group of high school students began publishing a local poetry magazine and the year I took Lewis Robinson's course. Since 1967, Talonbooks Publishing Company, founded by David Robinson and Jim Brown as a literary press, has published over 40 Canadian plays as well as volumes of poetry and fiction. Lewis Robinson's wife, Jo, has assisted the press as proof reader, while Lewis Robinson has published texts through Talonbooks, the most recent (1983) being *Concepts and Themes in the Regional Geography of Canada*.

The material, as distinct from the methods, used in this chapter closely resembles that pertaining to the Canadian imagination produced by Talonbooks and scores of other Canadian publishers. I hope by the end to assess not only the association between imagination and place, but also to assess the presence or absence of regionalism in self-conscious Canadian fiction and poetry. I invite you to judge these imaginations with me. Throughout we shall be asking ourselves whether or not we have embarked on an entirely subjective enterprise, an unsystematic look at writing. How do we choose the work to be assessed? How do we bring objective judgment to bear on the work? Fiction and poetry are fleeting and ephemeral when compared to other material commonly worked in the geographer's craft--distance, dollars, downtowns, directions, developments, dams and disasters.

Examinations of Canadian imagination has not matched, and likely will not match the bravado of geographers. The writing of Ronald Rees (1977) on imagination in the Prairies as recorded in literary and visual art is unique. The essay by Wreford Watson (1965) on regionalism and Canadian literature is another exception to familiar paradigms. But can the imagination of one important region of Canada be assessed outside the context of other regions? Is attention to the close associations among words used in the literature of each region sufficient? Perhaps. Is this, however, all we can say as geographers without resorting to opinions on the literatures of regions we have not experienced, or on writer's intended meanings? Yes or No?

Matters of imagination affect large numbers of people, and increasingly, a large number of writers as well by involving large groups of facts, changing theories of literature, and myriad feelings. A broad scope, however, need not produce an assessment based on bare generalities, private opinions, and favorite writers. We must first assume that writers whose works gain popular support, whose works are read by the public at large, and whose works are discussed in critical journals and published anthologies, use representative images and symbols of Canadians. To illustrate aspects of imagination expressed in literature, my task is to select passages of Canadian fiction and poetry from anthologies and works of criticism for analysis in time and space, and for comparison and interpretation. This will show how writers use metaphors of the physical environment, that is, how they use geography. Because a complete description is impossible, I shall supply only illustrations from publications of literary critics and writers of anthologies drawn from national journals and major books.

This essay also assumes that all poetry and fiction--the products of human imagination--has a logic whose essence may be found in geography. This supposition is perilous when applied to a country as large and as young as Canada, however, because its geographic milieux are diversified and underrepresented in the poetry and fiction of some regions; generalizations may promote confusion. Guides do exist for those wishing to trace the uses of geography in poetry and fiction. Some ground rules for poets and novelists are given by Robert Scholes and Robert Kellogg (1966) in *The Nature of Narrative*, a text telling much about resources and procedures employed in writing. Scholes and Kellogg not only describe writing traditions--varieties of romanticism, naturalism, realism and existentialism--they also define procedures for creating meaning, and for developing characters. Their assessments of describing and writing in the English language are based on a general point of view. The flowering of books and essays by Canadian authors is perhaps of greater importance to this subject because they, in the spirit of World War II, the Canadian centennial, the Constitutional debates, and the economic nationalism of the Federal Government, have become publicly introspective. Distinguished writers like Northrop Frye, Margaret Atwood, Desmond Pacey, George Woodcock, and Jack Hodgins to name a few have consciously tried to outline the ground rules comprising the traditions of Canadian poets and novelists to provide a well-spring of interpretation and bring forth the logic of literary geography.

Finally, how should we treat the problem of geographic fallacy? The supposition is absurd that we can find the authentic Canadian imagination by merely listing the elements of geography with which Canadian poets and novelists identify. When we uncover every word and image of geography used by Canadian writers, when we discover the richness of environmental imagery, we still will not have revealed the identity of Canadian writing. The truth of course is that, depending on the theory writers employ--romanticism or existentialism, for instance--environmental images are found in all poetry and novels. Their importance depends on the traditions followed. Because naturalism is the initial tradition of Canadian literature, we would be discouraged indeed to find that our literature is "placeless". We must not, however, fallaciously assume that Canadians are unique in the attention they give to geography. On the other hand, stress on the particular choice of imagery used to portray universal human situations, is authentic, and indeed is one of the purposes of this chapter. Nevertheless, a more important purpose is to show how the changing social milieux of Canadian life influence the ways Canadian writers imagine their relations to physical geography. Finally, I also commend to your attention the idea that literary place need not be isolated from historical and political realities. Literary place is but a moment in cultural history, although literary places of Canada are frozen in ice, or so we are told.

WORLD WAR II LITERATURE AND GEOGRAPHY

In 1943, Canada was at war. A.J.M. Smith (1943) in Ann Arbor Michigan was finishing the first modern criticism of Canadian literature, *The Book of Canadian Poetry*. Not far away, across the Detroit River, Lewis Robinson (1943) was beginning his career with a geographical paper on the city he knew so well, Windsor, Ontario. National unity was a central concern in all aspects of the nation, especially in political thought and in literary art. A.J.M. Smith's criticism portraying an international base of Canadian poetry has to be viewed in the context of the emergence of modernism, the cosmopolitan poets of the "roaring twenties," and the Depression. Except for the last sentence of his analysis, the war around him was totally ignored (Smith, 1943: 31). A connection between *The Book of Canadian Poetry* and national patriotism spurred by the war can be inferred from the principal review of the book by Northrop Frye who immediately noted a unity of tone in all the verse, a concept apparently avoided by A.J.M. Smith. Northrop Frye (1971: 138) commented on the definable Canadian genius:

> When all the intelligence, morality reverence and simian cunning of man confronts a sphinx-like riddle of the indefinite like the Canadian winter, the man seems as helpless as a trapped mink and as lonely as a loon.

To persuade us of this imagined national nature in contrast to the dominant metaphors of nature in other English-speaking nations, Northrop Frye (1971: 139) relied on Wilfred Campbell:

> Lands that loom like spectres while regions of winter, Wastes of desolate woods, deserts of water and shore; A world of winter and death, within these regions who enter, Lost to summer and life go return no more.

Although complying with the social milieu of nationalism and pride of the Canadian war effort, embodiment of related emotions in the geographical elements of climate, winter and empty spaces far exceeds our previous observations on geographic fallacy in poetry and novels.

In light of A.J.M. Smith's noncommittal attitude toward a unity in the imagery of Canadian nature, we can speculate on what he might have said about the novels appearing in the early war years. Canadian novels of this period show that the diversity of imagery observed in the years before the war did not last. Neil Macrae, the hero of Hugh MacLennan's (1941: 79) novel *Barometer Rising*, imagined a Canada stretching in winter white from night in Nova Scotia to noon in British Columbia:

> The sun had rolled on beyond Nova Scotia into the west. Now it was setting over Montreal and sending the shadow of the mountain deep into the

> valleys of Sherbrooke Street and Peel; it was turning the frozen St. Lawrence crimson and lining it with the blue shadows of the trees and buildings along its banks, while all the time the deep water poured seaward under the ice draining off the Great Lakes into the Atlantic. Now the prairies were endless plains of glittering, bluish snow over which the wind passed in a firm and continuous flux, packing the drifts down hard over the wheat seeds frozen into the alluvial earth. Now in the Rockies the peaks were gleaming obelisks in the mid-afternoon. The railway line, the tenuous thread which bound Canada to both the great oceans and made her a nation, lay with one end in the darkness of Nova Scotia and the other in the flush of a British Columbian noon.

Hugh MacLennan's (1945: 319) description of Canada in his more mature work, *Two Solitudes*, extended this earlier vision to encompass the north:

> In that autumn of 1939 the countryside in Canada had never seemed more tranquil. There was golden weather. In Nova Scotia and New Brunswick the moose came out of the forests on October nights and stood in silhouette against the moonpaths that crossed solitary lakes. In Ontario people looked across the water from their old river-towns, and seeing the lights of moving cars in the United States, remembered again that they lived on a frontier that was more a link than a division. On the prairies the combines rolled up the wheat, increasing the surplus in the granaries until it was hard to believe there were enough human mouths in the world to eat it all. In British Columbia the logs came down the rivers; people separated by mountains, plains and an ocean remembered English hamlets, pictured them under bombs, themselves islanded between snow-peaks and the Pacific. The Saint Lawrence, flowing past the old parishes, enfolding the Ile d'Orleans and broadening out in the sweep to Tadoussac, passed in sight of forests that flamed with the autumn of 1939; scarlet of rock maples, gold of beeches, heavy green of spruce and fir. Only in the far north on the tundra was the usual process of life abruptly fractured. Prospectors hearing on their portable radios that the world they had left was at war, could stand the solitude no longer; they broke

> camp, walked or paddled hundreds of miles southward, were flown out by bushpilots, appeared before recruiting stations in Edmonton, Battleford, Brandon, in the nearest organized towns they could find, and faded into the army.

MacLennan's landscape imagery devised from the regional diversity of Canada, although containing elements of idealism and patriotism not surprising to read at the end of a war, appeals both to geographers and to members of society at large; for every Canadian service man and woman who imagined Canada as being embodied in his or her Maritime or Prairie homeland, there were at least twenty if not fifty others just as idealistic as MacLennan serving in national institutions across Canada.

Canadian poetry of the 1940s may also be shown to express national unity. In his 1946 essay, "The Narrative Tradition in English-Canadian Poetry," Northrop Frye (1971c: 46) saw in Canadian poetry:

> A feeling of the melancholy of a thinly-settled country under a bleak northern sky, of the terrible isolation of the creative mind in such a country of resigning oneself to hardship and loneliness as the only means of attaining, if not serenity, at least a kind of rigid calm.

While he demonstrated this argument with quotations from east to west, from Isabella Crawford's "Malcolm's Katie," to Earl Birney's "David," the strongest support was taken from the Maritimes and the Great Lakes-St. Lawrence Lowlands. Northrop Frye (1971c: 153) relied most heavily on E.J. Pratt's *Brébeuf* with its greater reliance on the psychological tradition, to demonstrate the Canadian traditions of monster themes, tragic resolution, and human, but barbarous, natives. Throughout this important essay, Northrop Frye avoided the question of a diversified geography and the presence of conflicting images of nature in British Columbian verse. His view of Canada at the time of World War II may be described, not unfairly, as strongly fixed by his literary place--the Great Lakes-St. Lawrence Lowlands and the Maritimes. His specific use (or abuse) of geography, however, was not brought into play by World War II but by the Canadian centennial.

CENTENNIAL CRITICISM: USES AND ABUSES OF GEOGRAPHY

Northrop Frye used Canadian geography in his important essay written on the eve of the Canadian centennial and Expo 1967. Anticipating these national celebrations, Professor Carl Klinck (1965) edited the *Literary History of Canada* and called on Northrop Frye to write its concluding chapter.

The literary traditions of Canada in this history were described from east to west. From these impressions Northrop Frye (1971b: 225) could say:

> We may get some approach to characterizing the way in which the Canadian imagination has developed in its literature. Small and isolated communities surrounded with a physical or psychological "frontier," separated from one another and from their American and British cultural sources: communities that provide all that their members have in the way of distinctively human values, and that are compelled to feel a great respect for the law and order that holds them together, yet confronted with a huge, unthinking, menacing, and formidable physical setting--such communities are bound to develop what we may provisionally call a garrison mentality. In the earliest maps of the country the only inhabited centres are forts, and that remains true of the cultural maps for a much later time.

In the years following the Canadian centennial, including the year of the application of the War Measures Act, the garrison thesis of Northrop Frye resurfaced in the form of a guidebook for Canadian literature, *Survival*, written by his student, Margaret Atwood (1972). She identified eleven key patterns which occurred often enough to make Canadian literature distinct from other literatures. These patterns, Atwood (1972: 13) claimed, give shape to our literature and are "a reflection of the national mind." The most central of these patterns and most capable of demonstration from east to west is survival from hostile land, climate and natives; survival from disasters; "grim" not just "bare" survival; stubborn persistence to live after the withdrawal of that order nurturing the French and British empires.

Our literature employs symbols of nature (Atwood, 1972: 69-86) which pay special attention to animals in a way that is clearly different from the British tradition in which animals speak English, reflect class structure, and are generally protrayed as human beings. Atwood claims that our animal metaphors are also different from those used in American literature whose tradition portrays animals as objects of hunts or associated with learning experiences. The focus of the American quest is the hunter and the learner, not the hunted animal. Atwood claims the commonest use of animals in Canadian literature is to protray humans as stoic, doomed victims concerned with survival. Her evidence is selected from eastern Canadian nineteenth-century nature writers such as Ernest Thompson Seton or Sir Charles Roberts and also from contemporary authors--Alden Nowlan, Irving Layton, and Graeme Gibson. Had she sought evidence from the animal stories of Bill Bissett or Roderick Haig-Brown, to name only two writers west of the Canadian Shield, or had she looked at Newfoundland, could she have found similar support? Probably not.

In fact, the more we examine the survival thesis, the more we question its validity as a thesis defining anything beyond the literature and the physical setting of the Canadian Shield and its adjacent regions--the Great Lakes-St. Lawrence Lowlands and the Prairies. The thesis gives little attention to variations in climate compared to variations in land and animals of Canadian fact and fiction. In her special appendix on snow, Atwood (1972: 65) said, "It is in their attitudes towards winter that Canadians reveal most fully their stance towards Nature--since winter for us is the 'real' season." The West Coast climate and a different aspect of the Canadian imagination here are ignored. Margaret Atwood might well discard such a focus in a retrospective glance a decade and a half after the centennial. For example, would she use geography differently in 1984? Is she a regional or a national critic? To someone who has written so much about "survival," regionalism in fact may be more attractive in the world of Nineteen Eighty-Four.

Perhaps the most valuable single criticism of Canadian literature stemming from the centennial commemorations, especially to those writing geography, is Desmond Pacey's (1970) " 'Summer's Heat and Winter's Frigid

Gales' : The Effects of the Canadian Climate Upon Canadian Literature." Although others have drawn attention to climate's effect, Desmond Pacey's work appears to be the most systematic and weighted evaluation. In Pacey's assessment, we read that Canadian poetry and novels from the beginning to the present day have been obsessed by appearances from climate and landscape at various seasons. The principal effects on our literature are "certain words, many images and symbols, a plethora of descriptive passages, a strong reliance on seasonal rhythms as a principal structure and finally certain attitudes to the human situation" (Pacey, 1970: 10).

That words and images in Canadian literature owe something to climate has never been questioned. We all recognize the special meanings of "forty below" (forty below zero), "whiteout" (sight problems in driving snow), "black ice" (paved road ice), "squaw winter" (mild winter in the beginning), "black blizzard" (dust storm), and "freshet" (streams swollen with melting ice). Many images refers to climate. Characters are "caught in snow storms of paper," "fall and break into snowflakes on cold stones," and live in "white regions of winter." Doom is evoked by Bliss Carmen (Pacey, 1970: 12) with climatic symbols:

> The windows of my room
> Are dark with bitter frost.

Death is similarly evoked by Earle Birney (Pacey, 1970: 13):

> Man is a snow that cracks
> the trees' red resinous arches
> and winters the cabined heart
> till the chilled nail shrinks in the wall
> and pistols the brittle air
> till frost like ferns of the world that is lost
> unfurls on the darkening window. . . .

Given the literature before him, Desmond Pacey had no difficulty in showing the climatic imagery in descriptive settings. He pointed to the abundant evidence offered across the nation: P.K. Page's "Stories of Snow," Alden Nowlan's, "Midnight of the First Snow," Dorothy Livesay's, "Winter," Irving Layton's, "First Snow,"E.J. Pratt's, November Landscape," Frederick Grove's, *Over Prairie Trails* , Morley Callaghan's, *The Loved and the Lost* , Mordecai Richler's, *Son of a Smaller Hero*, or Al Purdy's, *Wild*

Grape Wine . Literary traditions in which climate affects word choice , metaphor, and setting, serve to define our literary place.

Climate's effect on structuring the Canadian novel is even more outstanding. Although the idea of structuring the novel by seasons is not solely Canadian, and is present in other literary traditions, especially those influenced by the naturalist movement, it is virtually universal in the works of Canadians--even in those of modern novelists working in the existentialist tradition. True not only of Frederick Grove and Ethel Wilson, it is part also of Morley Callaghan and Scott Symons (1967) (*Place d'Armes*).

When Desmond Pacey looked at human situations revealed by climatic metaphors, he tightly qualified Northrop Frye's garrison thesis. Using Canadian geography with respect, and finding evidence for more than the philosophy of man's struggle with a hostile climate, Pacey (1970: 15) concluded:

> Just as our climate swings widely between extremes of heat or cold so our literature swings from dream to nightmare and back again. On the one hand is a constant, frightening sense of man's fragility on the other an exhilarating sense of his capacity to endure, to transform, to comprehend.

A nation's centennial celebrates nationalism. If nationalism, as distinct from regionalism, has never produced profound literature, nationalism is also unlikely to produce profound criticism. In the Canadian centennial, critics both used and abused Canadian geography. By conceiving Canadian poetry and novels as showing one overwhelming response or even several overwhelming responses to the hostility, especially the winter,of Canadian geography, some of the centennial's literary criticism presented the entirely incorrect idea that Canada's physical setting is almost uniform. This is obviously absurd to anyone who looks at a climatic map or who travels, for example, from other parts of Canada to British Columbia.

One critique of Canadian writing as an expression of human imagination, written in the centennial decade by a geographer, portrayed Canadian unity in geographical and literary diversity. In "Canadian

Regionalism in Life and Letters," Wreford Watson (1965) used virtually the same literary sources as Desmond Pacey, Margaret Atwood, and Northrop Frye. Relying heaving on direct quotation, Watson (1965: 214) concluded that regional individuality and regional problems breed regional minds and thought:

> But regionalism need not mean disintegration. It may mean diversification. And as long as all the diversities involved recognize their dependence upon each other, they can actually add to the idea of unity. Certainly they can enrich that unity by giving it many ways of expression.

I clearly can see Lewis Robinson nodding his head in support of this conclusion.

REGIONALISM AND THE LOTUS THEME OF BRITISH COLUMBIA: 1982

In the decade following the centennial and declaration of the War Measures Act, literary criticism changed direction. Cues to the new alignment appeared in the editorial policies of Canadian publishing houses. Macmillan Company devised a policy whereby critics across Canada assessed the themes of Canadian literature. Of the sixteen publications appearing in the series by 1976, seven assessed the question of regionalism, not national unity. Michael Nowlan (1975) edited *The Maritime Experience*, Terry Angus (1975) *The Prairie Experience*, and John Stevens (1975) *The Ontario Experience*; Quebec, the North, and the West were given similar treatment. In British Columbia, *Canadian Literature* produced a special 1976 issue on the Maritimes, one of several special issues published on regional literature in Canada during the years of national debates on the constitution and on establishment of a new energy policy to restructure tariffs and to promote the growth of the energy-rich West.[2] Some observers argue that criticism of Canadian poetry and novels written in this social milieu outside the Great Lakes-St. Lawrence Lowlands introduced separatist polemics into Canadian writing. Even if major editors and writers were to advocate separatism, on what conceivable grounds could they destroy the garrison or survival thesis?

By inspecting more closely the assessment in one region, British Columbia, we can comprehend some of the relationships between regionalism and the established traditions of Canadian literature. The editorial pages of *Canadian Literature* contained some of the earliest qualifications of the garrison and survival theses. Responding to mild counter claims, George Woodcock (1973: 5) wrote, "yet considered as a critical tool, a frame of reference, a usable hypothesis it is hard not to accept Atwood's horizon of survival as the circle that best defines the bounds of Canadian writing." When the British Columbian author, Jack Hodgins (1975; 1976), was asked to edit a series of themes in Canadian literature by Macmillan, he found it possible to be explicit in the uses of geography by British Columbian writers. Quoting often from some of the authors cited by Margaret Atwood, Hodgins concluded that, although the literary place was mapped with mountains, salt water, brooding cliffs, and crashing waves, maps are drawn differently by those in the literary domain of the "garrison," and those who merely survive. Jack Hodgins selected poems and parts of novels to show that the British Columbian imagination dwells on a caring and loving relationship with a benevolent nature, a nature sufficiently promising to inspire a greater caring for other people. In British Columbian imagination, geography is not used as a metaphor for hostility but for heaven-on-earth, or Eden. "There is," Jack Hodgins (1976: 2) said, "a sense here of living at the edge of the continent, that this is also the extreme frontier of the world. Old realities can be thought of as left behind in old countries (England or Hungary or Nova Scotia or Ontario or wherever) and new realities take their place, the realities of Lotus Eaters perhaps." Many, even those from the Great Lakes-St. Lawrence Lowlands, concur: "That is what happened to my imagination in this landscape."

In a decade rich for Canadian regionalism, two criticisms about the literary traditions of British Columbia emerged in sharp contrast with the garrison-survival view of Canadian literature. In the criticisms of Jack Hodgins (1975; 1976) and Allan Pritchard (1982), the uses of geography made by British Columbian writers are the very opposite of those described by the Great Lakes-St. Lawrence critics.

Although similar to Northrop Frye and Margaret Atwood in his analysis of how metaphors are drawn from the natural world, landscape, and climate, Allan Pritchard has shown us a view of man predominantly celebrating the landscape. In the Edenic qualities of gardens, pastures, seas and forests of British Columbia, man finds in nature perfect happiness, not alienation and disappointment. The imagery of nature used by Martin Allerdale Grainger (1908) in *Woodsmen of the West* portrays man taking to the countryside "as sunflower takes to the sun" (Pritchard, 1982: 99). In Theodora Stanwell-Fletcher's (1946), *Driftwood Valley,* we read of settlers finding perfect happiness in the interior plateaus of the Cordillera. The images of Ethel Wilson's (1949) *The Innocent Traveller,* Cliff Kopas'(1976) *Packhorses to the Pacific,* and the works of Roderick Haig-Brown, Malcolm Lowry and Jack Hodgins strikingly contrast with those described by Northrop Frye and Margaret Atwood. In the essay collections of Roderick Haig-Brown (1946-1950), *A River Never Sleeps,* and *Measure of the Year ,* nature is a pleasure intimately to know, to possess and to conserve, not a monster to face with trepidation. Roderick Haig-Brown gave British Columbia the symbol of itself as Lotus Land. Neither he nor Malcolm Lowry showed a fascination with winter. Spring is the season that fires the imagination of British Columbian writers (Pritchard, 1982: 108). Whether the theoretical bent is Ethel Wilson's romantic naturalism, Malcolm Lowry's "art deco" surrealism, or Jack Hodgin's neo-realism, British Columbia's imagination celebrates an earthly paradise; it is spring not winter.

CONCLUSIONS

Many literary critics recognize that the use of geography is an important element of Canadian poetry and novels. When Lewis Robinson began his career in writing and teaching geography, Canadian literary criticism was preoccupied with describing from east to west a physically and culturally diversified landscape that embodied the spirit of a nation. The transcontinental geographical image of Canada used in the decade of World War II was similar to that emphasized by the Confederation writers. It was somewhat similar to that found in the literary criticism associated with commemoration of the Canadian centennial. Ontario critics for example, originally described Canada on the basis of their literary place within its geography--thinly populated Canadian shield, inhospitable climate, and hostile Indians. They deduced from this regional geography the national patterns of Canadian literature but ignored the diversity of Canadian life and literature. A strikingly different Canadian imagination was defined later through such works as the cosmopolitan and psychological introspection of A.J.M. Smith (1943). Recent critics have perceived quite divergent images of geography being used in regions such as British Columbia which cannot be understood without probing the sentiments expressed in Western Canadian opposition to the National Energy Policy and the Canadian Constitution (Gibbins, 1983).

We cannot assume that Canadian poetry and novels are distinctive, however, because they use geography. Geography and especially climate provide distinctive metaphors in the literature of any English-speaking nation. Geographers' spiritual great grandfather, Charles Darwin, the modern architect of naturalism, is also the great grandfather of nineteenth century English literature. Some would say that Charles Darwin is the architect of Canadian intellectual thought as a whole, much as John Locke is the architect of intellectual thought in the United States. The consequence of naturalism in all English-speaking literature, including the literature of the tropical United States and Australia (Brydon, 1981), is that writers use geography. To say that Canadian metaphors focus on snow, those in Australia on sun, is not to say much: all metaphors, all poetry, and all novels are derived from immediate experience. At least since Darwin, that experience is essentially regional. Even the surrealistic images of modern poetry contain literary "places."

This is not to say that geography determines metaphor. Not the physical but the social milieu influences "how" we use geography in literature. The use of geography in the making of Canadian literature depends on the moment of Canadian history; geography is used differently in moments of external military threat and to commemorate important events of our political existence than in moments dominated by domestic threats to material well-being, economic prosperity and political freedoms such as for example, those exposed in national television debates.

The uses that English-Canadian poets and novelists make of geography are not in any direct way related to Lewis Robinson's geography of Canada, a lifetime devoted to careful observation, measurement, and classification of Canadian land and life eschewing personal images of the country in favor of those which can be recorded only by the concrete data of mental maps. Nevertheless, observations recorded by the literary imagination in Canada and associated with the country's physical milieux became personally attractive as a result of Lewis Robinson's teaching and through images of the whole man-geographer in the mirror he held up for me. Escape is impossible from aspects of the imaginary parts of our life--the visual and literary arts of the nation--when teaching the geography of Canada. These artistic images fuse with Lewis Robinson's regional geography of Canada and permit us all to learn through him how to answer the question, "Where is here?"

NOTES

1. I wish to acknowledge assistance from the Geography Department, California State University, Fresno, in preparation of this manuscript.

2. Special Issue on the Maritimes, (1976), *Canadian Literature*, 68-69: 3-152.

REFERENCES

ANGUS, T. (editor) (1975), *The Prairie Experience* (Toronto: Macmillan, Themes in Canadian Literature).

ATWOOD, M. (1972), *Survival* (Toronto: Anansi).

BRYDON, D. (1981), Landscape and authenticity: The development of national literatures in Canada and Australia, *Dalhousie Review*, 61: 278-90.

CHAMBERS, R.D. Notes on regionalism in modern Canadian fiction, *Journal of Canadian Studies,* 11: 27-34.

FRYE, N. (1971a), Canada and its poetry, *The Bush Garden* (Toronto: Anansi): 129-143.

FRYE, N. (1971b), Conclusion to a literary history of Canada, *The Bush Garden. Essays on the Canadian Imagination* (Toronto: Anansi): 213-251.

FRYE, N. (1971c), The narrative tradition in English-Canadian poetry, *The Bush Garden* (Toronto: Anansi): 145-155.

GIBBINS, R. (1983), Regional politics in the wake of the Canada Act, *Canadian Issues,* 5: 131-137.

GRAINGER, M.A. (1908), *Woodsmen of the West* (Toronto: Musson).

HAIG-BROWN, R. (1946), *A River Never Sleeps* (Toronto: Collins).

HAIG-BROWN, R. (1950), *Measure of the Year* (Toronto: Collins).

HODGINS, J. (editor) (1975), *The Frontier Experience* (Toronto: Macmillan, Themes in Canadian Literature).

HODGINS, J. (editor) (1976), *The West Coast Experience* (Toronto: Macmillan, Themes in Canadian Literature).

KLINCK, C.F. (editor) (1965), *Literary History of Canada* (Toronto: University of Toronto Press).

KOPAS, C. (1976), *Packhorses to the Pacific* (Sidney, B.C.: Gray's).

MACLENNAN, H. (1941), *Barometer Rising* (Toronto: McClelland and Stewart, New Canadian Library).

MACLENNAN, H. (1945), *Two Solitudes* (Toronto: Collins, White Circle Pocket).

NOWLAN, M. (editor) (1975), *The Maritime Experience* (Toronto: Macmillan, Themes in Canadian Literature).

PACEY, D. (1970), "Summer's heat, and winter's frigid gales": The effects of the Canadian climate upon Canadian literature, *Royal Society of Canada. Proceedings and Transactions,* 4th Series, 8: 3-23.

PRITCHARD, A. (1982), West of the great divide, *Canadian Literature,* 94: 96-112.

REES, R. (1977), Prairie: A Canadian artist's view, *Landscape,* 21: 26-32.

ROBINSON, J.L. (1943), Windsor, Ontario, a study in urban geography *Canadian Geographical Journal*, 27: 106-121.

ROBINSON, J.L. (1983), *Concepts and Themes in the Regional Geography of Canada* (Vancouver: Talonbooks).

SCHOLES, R. and KELLOGG, R. (1966), *The Nature of Narrative* (New York: Oxford University Press).

SMITH, A.J.M. (editor) (1943), *The Book of Canadian Poetry* (Toronto: W.J. Gage and Co,).

STANWELL—FLETCHER, T. (1946), *Driftwood Valley* (Boston: Little Brown, and Co.).

STEVENS, J. (editor) (1975), *The Ontario Experience* (Toronto: Macmillan, Themes in Canadian Literature).

SYMONS, S. (1967), *Place d'Armes* (Toronto: McClelland and Stewart).

WATSON, W. (1965) Canadian regionalism in life and letters, *Geographical Journal*, 131: 21-33.

WILSON, E. (1949), *The Innocent Traveller* (Toronto: Macmillan).

WOODCOCK, G. (1973), Horizon of survival, *Canadian Literature*, 55: 3-6.

With Santa, at an Undergraduate Christmas Party, 1979

Hockey game with Geography Graduate Students, 1981
(Despite a heart attack which caused Lewis Robinson to resign in 1968 as Head of the Geography Department, he has played Friday afternoon hockey with the department's graduate students every year since 1972)

THE REGIONAL DISTINCTIVENESS OF CANADIAN CITIES

Charles N. Forward

The distinctiveness of the major regions of Canada has been emphasized by many geographers who customarily divide the country into regions for analysis of specific areas and subsets of phenomena. Canadian cities, on the other hand, are usually studied as a total system, with regional characteristics sometimes being discovered as significant findings of the research. The major goal of this essay is to produce a comprehensive characterization of the cities within each of the widely accepted major Canadian regions: the Atlantic Provinces, Quebec, Ontario, the Prairie Provinces, and British Columbia. The study takes into account those elements of urban character that are not easily measured, and those that can be inferred from census data. The presentation here, however limited by space to a preliminary sampling of the potential evidence, briefly sketches the regional distinctiveness of Canadian cities, and concludes with a call for a more complete investigation.

THE COMPARATIVE STUDY OF CANADIAN CITIES

Examining the study of Canadian cities by geographers, Simmons discovered that, "until the mid 1950s the only studies in print were a few profiles of individual cities, the work of regional geographers momentarily turning their attention to an urban area" (Simmons, 1967: 341). Nevertheless, the roots of urban geography in Canada date back to the late 1940s when Griffith Taylor published his text book, *Urban Geography* (Taylor, 1949) which was one of the first to focus on the subject, and predates the well known volume of Mayer and Kohn (Mayer and Kohn, 1959) by ten years. Taylor's was written in Canada and its comments were illustrated with many Canadian examples.

Taylor's *Urban Geography* focused on:

(1) prehistoric settlements and marginal survivals, using the zones and strata theory as a framework for understanding distributions;

(2) the evolution of town plans;

(3) the site factor in the evolution of towns; and

(4) the seven ages of towns (stages in evolution).

These themes had remarkably little influence on the development of urban geography in Canada and were largely ignored by most urban geographers who first became infatuated with the revival of Christaller's central place theory and then joined the quantitative revolution. The ability to interpret topographical and geological influences on urban sites and development, which Taylor so admirably practiced, regrettably has been lost to urban geography. Unlike the abstract entities discussed in many of the later statistical studies, Taylor's cities were real and rooted firmly in the natural environment.

The first studies to deal with the whole city system of Canada appeared in the mid-1960s. Maxwell's (Maxwell, 1965) functional classification represented an application to Canadian cities of the minimum requirements method of determining the urban economic base (Ullmann and Dacey, 1960). Basing the article on his Master's thesis executed under the supervision of J. Lewis Robinson at the University of British Columbia, Maxwell produced the first significant regionalization of Canadian cities. He distinguished between heartland and periphery cities, and further subdivided the periphery into eastern, northern and western sections. Several later studies of functional urban characteristics based on subsequent censuses utilized the same approach and general methodology (Maxwell, Greig and Meyer, 1972; Li, 1976; Li, Scorrar and Williams, 1978). Kerr also considered the hinterland concept, especially the relationship between Toronto and the hinterland cities, in his study of the geography of finance in Canada; his chief concern however, was the significance of cities to the realms of finance and insurance (Kerr, 1965).

King's analysis of urban dimensions represented the full flowering of the quantitative analysis of Canadian cities; it used factor and discriminatory analysis to handle fifty to sixty variables instead of Maxwell's seven or eight (King, 1966). Although the leading variables were based on census data concerning demographic, housing and labor force characteristics, a number of others assessed cities according to age, "age of city," and transportation facilities, "served by commercial airlines," "having port facilities." Many of king's city groupings, while not comprised entirely of cities from a single region, were dominantly regional in character. City groups were clearly derived for Ontario, Quebec and the industrial frontier (King, 1966: 220).

Although many studies of Canadian cities have appeared since the 1960s, most have been concerned with special aspects of urban areas and have not been oriented to regions per se. Ray's illuminating study of Canadian cities is a major exception to this trend (Ray, 1971). Based on an extensive analysis of a number of economic, demographic and cultural variables listed in the 1961 Census of Canada, Ray identified French and English cities, the British Columbia city, Maritime-Prairie city contrasts, and general distinctions between cities of the Center and those of the Periphery.

Preston's study of the evolution of urban Canada since 1867 was organized regionally in part (Preston, 1978) under the headings, Central Canada, Atlantic Region, Prairie Region and British Columbia. In his study of the Canadian urban system, Simmons clarified the nature of the urban hierarchy, indicating clearly that below the fifth order centers of Toronto and Montreal there were regional hierarchies focusing on fourth order centers, such as Halifax, Winnipeg and Vancouver (Simmons, 1976a). Many other studies have focuses on cities in particular regions of Canada, but have not undertaken comparisons with other city groups in the nation (Bourne and MacKinnon, 1972; Forward, 1972; Hardwick, 1972; Marshall, 1972; Smith, 1972a, Bourne _et al_., 1974).

Some studies of urban functional characteristics and industrial diversification have elaborated themes initially investigated by other writers or proposed in different contexts. Marshall (1975; 1981) has pursued themes previously studied by Maxwell. Special aspects analyzed for Canadian cities include migration (Simmons, 1977), incomes (Simmons, 1976b; Forward, 1982), nodal structures based on newspaper news datelines (Kariel and Welling, 1977; Kariel and Rosenvall, 1978) and metropolitan dominance and foreign ownership (Semple and Smith, 1981). Most of these studies have thrown new light on the nature of the urban hierarchy and help to differentiate cities on a regional basis.

Very little however, has been written by geographers on the character or personality of Canadian cities. One of the few discussions of this subject is in Simmons and Simmons (1974: 27-44). Another is by Mercer (1979) whose comparison of Canadian and American cities concluded that the former were distinctive and perhaps should not be described as peas in a pod with American cities in text books on "the North American city." On the basis of this distinction between Canadian and American cities, one can take a step further and formally recognize a distinctiveness of cities within the major regions of Canada.

REGIONAL DISTINCTIVENESS

The lack of universal agreement on a set of geographically distinctive major regions of Canada means that relevant criteria must be chosen. Agreement is more common that the Atlantic Provinces, the Prairies, British Columbia and the Territories are separate regions than that Ontario is distinct from Quebec. On the grounds of heartland and even on physical characteristics (Great Lakes-St. Lawrence Lowlands) many geographers consider southern Ontario and Quebec as a single major region, and place the rest of those provinces in the Canadian Shield, a region which may also include Labrador and the Shield portions of the Prairies (Warkentin, 1968; McCann, 1982; Robinson, 1983). The great significance of the French language and culture in Quebec cities however, serves to enhance their distinctiveness from Ontario cities. Recognizing the geographical imperfections of political regions, this study uses the divisions adopted for the International Geographical Union monographs on Canadian regions (Gentilcore, 1972; Grenier, 1972; Macpherson, 1972; Robinson, 1972; Smith, 1972b; Wonders, 1972). The Territories were omitted from consideration because only one city, Whitehorse, has a population greater than 10,000.

Characterizations were based on the criteria of size, age, setting and ethnic background mentioned by Simmons and Simmons (1974: 27-44) as well as other factors, such as climate, architecture, economic function, language, mobility, education, family status and housing. Information was derived from the literature, statistics and personal observations. Some tabular presentations in this study are based on 1971 census data processed previously by Ray to indicate various statistical dimensions of Canadian cities (Ray, 1976). Relevant data from 1981 census were not yet available when this study was carried out. Because the following strictly are generalizations, some, or even many, individual cities may not exhibit all components of the urban character attributed to their respective regions.

ATLANTIC PROVINCES

Cities in the Atlantic Provinces tend to be smaller and slower growing than those of other regions. The level of urbanization is low and the region has been by-passed by immigration for much of the past century. The fragmentation of the region into a number of islands and peninsulas has produced many rather limited hinterland areas in which a number of central places have arisen; the resulting transportation and communication impediments to interaction, however, have prevented any city from achieving urban dominance. Although Halifax is the largest city, its dominance is not strongly developed. A maritime orientation is almost universal; because most of the cities are coastal they have developed as ports. Fishing and shipping activities are characteristic urban functions. The ever-present influence of the ocean has moderated the humid continental climate. The cities are not so warm in summer and not so cold in winter as those inland, and their snow depth is not as great. Although grid patterns of streets are common, irregularity of layout has often been imposed by sloping land and the shapes of harbors. The perennial low economic growth of the Atlantic Provinces has been reflected in the generally old appearance of the region's cities. Central business districts tend to have fewer new buildings and more heritage structures than those of other regions, and the cities consequently project more of an old-world, historic character. The older houses are distinctive architecturally--two or three story wood-frame, clapboard structures in which verticality is emphasized by gently sloping roof lines. Many are row houses and have common side walls.

Various population characteristics emphasize regional distinctiveness. Atlantic cities have youthful populations comparable with those of cities in Quebec (Table 1). Both regions have distinctly lower life cycle indexes than others in Canada; the Atlantic cities have the highest proportion of children, as well as the highest birth rate. Families tend to be larger, as in Quebec, but in educational levels achieved, Atlantic cities rank well ahead of Quebec and are similar to those in Ontario and the Prairies. The Atlantic cities are the most British in character, as evidenced by the ethnic origin and mother tonque statistics, but also register the strongest French presence outside Quebec (Table 2). Other ethnic groups and mother tongues are sparsely represented, producing a low index of ethnic diversity. Not only is religious diversity low compared with the rest of English-speaking Canada but the Atlantic cities are less Protestant and more Catholic. Figures on birthplace and migration indicate that the stable population composed of the founding peoples was mostly born in Canada and is not characterized by great mobility (Table 3).

Housing and economic functions derived from the Canadian census also help to identify the region's character. Apartments constitute one-quarter of the housing stock, a proportion similar to that of Ontario and the Prairies (Table 4).

TABLE 1

AGE, SEX, BIRTH, DEATH, FAMILY AND EDUCATION CHARACTERISTICS, 1971

Characteristics	Atlantic	Quebec	Ontario	Prairies	British Columbia
Number of Cities	20	34	45	18	19
Age Groups (per cent)					
0-14	32	30	29	29	31
15-64	61	64	62	62	61
65 and over	7	6	9	9	8
Life Cycle Index[1]	78	79	98	96	89
Males per 100 Females	98.6	98.4	98.3	99.4	103.5
Birth Rate per 1,000	20.7	16.1	16.8	19.8	19.1
Death Rate per 1,000	7.9	6.5	8.2	8.2	8.0
Average Family Size[2]	4.0	4.0	3.5	3.6	3.6
Education: Highest Level Attended (per cent)					
Elementary	37	51	37	35	23
Secondary	54	43	55	55	65
Some University	5	3	4	5	8
University Degree	4	3	4	5	4

[1]Life Cycle Index = $\frac{\text{Population aged 45 and over}}{\text{Population aged 0-14}}$ (the lower the value, the higher the proportion of younger people).

[2]The census family consists of husband and wife, plus unmarried children (of any age) living at home.

TABLE 2

ETHNIC, MOTHER TONGUE AND RELIGIOUS CHARACTERISTICS, 1971

Characteristics	Atlantic	Quebec	Ontario	Prairies	British Columbia
Number of Cities	20	34	45	18	19
Ethnic Origin (per cent)					
British	74	6	61	49	56
French	19	91	15	6	6
Other	7	3	24	45	38
Index of Ethnic Diversity[1]	.31	.16	.53	.70	.64
Mother Tongue (per cent)					
English	84	6	80	80	83
French	14	93	10	3	2
Other	2	1	10	17	15
Religious Denomination (per cent)					
Roman Catholic	47	94	36	24	22
Principal Protestant[2]	38	4	48	53	53
Other	15	2	16	23	25
Index of Religious Diversity[3]	.64	.11	.74	.81	.81

[1]Index of Ethnic diversity = $1 - \Sigma pi^2$, where pi = the proportion of an urban area's population in the ith ethnic group. Twelve ethnic groups were used, the twelfth being a residual "other" category. The higher the value, the more ethnically diverse is the population.

[2]Principal protestant includes Anglican, Lutheran, Presbyterian, and United.

[3]The index of religious diversity was derived in the same manner as the index of ethnic diversity. It was based on twelve groups, one of which was an "other" category. The higher the value, the more religiously diverse is the population.

TABLE 3

BIRTHPLACE AND MIGRATION CHARACTERISTICS, 1971

Characteristics	Atlantic	Quebec	Ontario	Prairies	British Columbia
Number of Cities	20	34	45	18	19
Birthplace of Population (per cent)					
Same Province	82	95	74	65	48
Different Province	14	3	11	20	32
Outside Canada	4	2	15	15	20
In-Migration Ratio[1]	21	17	21	29	34
Out-Migration Ratio[2]	18	17	17	23	27
In-Migrants from Within Canada (per cent)	92	95	83	89	85

[1]The in-migration ratio is the ratio of in-migrants to the 1971 base population where both numerator and demonimator pertain to persons aged five and over in 1971.

[2]In the out-migration ratio the numerator is the number of persons who resided in Canada outside the pertinent area on June 1, 1971, but who had resided within the area on June 1, 1966.

TABLE 4

HOUSING CHARACTERISTICS, 1971

Characteristics	Atlantic	Quebec	Ontario	Prairies	British Columbia
Number of Cities	20	34	45	18	19
Apartments (per cent)	25	39	25	24	19
Dwellings Rented (per cent)	40	49	35	39	34
Median House Value ($)	15,000	15,000	19,000	17,000	20,000
Average Cash Rent ($)	98	80	108	104	111
Average Family Income ($)	8,579	8,846	10,057	9,401	9,934
Rooms per Dwelling	5.7	5.2	5.6	5.3	5.3
Persons per Room	.68	.74	.60	.62	.62

A resemblance to Quebec cities is observed in measures relating to prices and living standards, such as house value, cash rent and family income, which tend to be lower than in Ontario and western cities. In the urban economies of Atlantic cities public administration and defence looms larger than in any of the other groups, while manufacturing is less important than anywhere else except on the Prairies (Table 5). Central place functions, represented by transportation, communication, trade and service activities, comprise thirty-five per cent of the dominant economic functions of the region's cities.

QUEBEC

The French language and culture differentiate Quebec cities from those of other Canadian regions. To a visitor from English-speaking Canada signs of all types and sizes loudly proclaim the French character, before the spoken word is heard. French communication networks and facilities, such as the Canadian Broadcasting Corporation radio and television networks, and the metropolitan newspapers, help to unify the urban system of Quebec. The overwhelmingly presence in the urban hierarchy of Montreal, and the existence of a prominent second city, Quebec, both much larger than the rest, offer a marked contrast with the situation in the Atlantic Provinces. The majority of Quebec cities, focused on the St. Lawrence River, Estuary and Gulf, are riverine or coastal in location; many are ports. The streets of most cities conform to a grid pattern modified by local variations in topography and irregularities caused by water bodies. Quebec cities tend to have colder winters and heavier snowfalls than those in either Atlantic or Ontario Canada, but summers are not quite so warm as those in Ontario. Old for the most part, many cities have historic areas that have been preserved, especially in Quebec. The abundance of large churches and associated religious institutions in Quebec cities attests to the traditional strength of Roman Catholicism. Brick and stone building materials are more common than in Atlantic cities, and architectural forms are distinctive among the older buildings. The steeply pitching roofs with dormer windows and the outside stairways

TABLE 5

DOMINANT ECONOMIC FUNCTION,[1] 1971

Characteristics	Atlantic	Quebec	Ontario	Prairies	British Columbia
Number of Cities	20	34	45	18	19
Industry	Per cent of Total Cities within Region				
Extraction	15	15	9	17	0
Manufacturing	25	64	73	11	42
Community and Personal Services	15	15	5	39	5
Public Administration and Defence	25	6	11	17	16
Other	20	0	2	16	37
Total	100	100	100	100	100

[1]The dominant functions were derived by the use of the minimum requirements method.

giving access to the second and third floors, characterize many multi-family dwellings.

The age structure of Quebec cities like that of Atlantic Canada is influenced significantly by youthful cohorts (Table 1). Birth rates, which historically were higher in Quebec than in most other provinces dropped during the 1960s, leaving cities in Quebec with the lowest rate among the five Canadian city groups by 1971. On the other hand, Quebec cities were also the lowest in death rates. The levels of education attained by people in Quebec cities, however, are different from those achieved elsewhere in urban Canada. Half the population of Quebec cities has obtained no more than elementary education; Quebec cities ranked well below the Canadian average in every category recording attainment of secondary and university education. The distinctiveness of Quebec cities is also protrayed dramatically in ethnic origin, mother tongue and religious denomination (Table 2). Only Montreal, with its large British and postwar immigrant minorities, contradicts the almost universal French-Catholic character of Quebec cities. In view of the linguistic isolation of Quebec within the nation, mobility understandably is low and most of the urban population was born in the province (Table 3).

The proportion of apartments in Quebec cities is higher than in other Canadian city groups, and the proportion of single family dwellings is correspondingly lower (Table 4). The percentage of tenancy is thus likely higher in Quebec. Average rents are by far the lowest, however, and the low average house values correspond to those of Atlantic cities. Standards of living indicated by average family income while a little higher in Quebec cities than in those of the Atlantic region, are significantly lower than elsewhere in the nation. Quebec cities also lead in persons per dwelling room and suggest that living conditions are more crowded in urban Quebec than elsewhere in Canada.

Another pronounced characteristic of Quebec cities is the role of manufacturing as the dominant economic function: nearly two-thirds of Quebec cities occupy this category (Table 5). The remarkably low percentage of Quebec cities oriented to public administration and defence, however, contrasts sharply with the large proportion recorded for this activity in Atlantic cities.

ONTARIO

Ontario cities focus on the Great Lakes and reflect a national spatial, economic and cultural centrality. The southward thrust of the province into the heart of the United States manufacturing belt has also facilitated interaction of Ontario and American cities. Ontario cities have a more pronounced southward orientation than that of cities in other groups. The primacy of Toronto in the Ontario urban hierarchy is comparable to that of Montreal in Quebec, but the Ontario hierarchy with six cities in the one-quarter to three-quarter million size category is more robust than the Quebec hierarchy which has only one such city. The Great Lakes' focus of Ontario urban places reflects the port function of many of the region's largest cities, and the general proximity to the Great Lakes and St. Lawrence Seaway of many others. Ontario cities have been more prosperous during the present century than those of the Atlantic Provinces or Quebec and frequently look younger than they really are as a result of extensive expansion and redevelopment. Streets usually conform to a grid pattern modified by local physical irregularities. Brick is the favorite material for residential construction and the style of older houses reflects the gothic revival form comprising sharply pitched roofs, multiple gables, and decorative woodwork on eaves and gable ends.

Many population characteristics of Ontario cities resemble more those of the two western regions than those of the eastern ones. The populations of Ontario cities are older than in the east, death rates are higher and family sizes are smaller (Table 1). In educational levels achieved by the population, however Ontario cities are almost the same as those in Atlantic Canada. Like them, Ontario cities also have a high component of population which is French by origin and mother tongue, but compared to Atlantic cities they have greater ethnic diversity--nearly one-quarter of the urban population is of non-British, non-French origin (Table 2). A large proportion of the postwar immigrants to Canada have gone to Ontario cities. Within English-speaking Canada, Ontario cities lie part way between the Atlantic and western cities in characteristics of religious denomination. They are less Catholic and more Protestant than in the east, but have less religious diversity than cities in the West. Place of birth of the population and proportion of in-migrants from within Canada suggest that Ontario cities have a more cosmopolitan population than those of the eastern regions--only eighty-three per cent of its immigrants are from within Canada, the lowest such percentage for all regions (Table 3). However, in-and out-migration ratios, and the proportion born in other provinces, are distinctly lower than in the West.

The people of Ontario cities are generally prosperous and well housed (Table 4). They have the highest average family income of all city regions and the lowest number of persons per room. Home ownership is high, comparable with that of cities in British Columbia, and apartment living, at twenty-five per cent, is no more prevalent than in Atlantic cities. House values and dwelling rentals are the second highest in Canada after those of cities in British Columbia.

The dominant economic function of Ontario cities is manufacturing; nearly three-quarter of the cities fall in this category (Table 5). Despite the extensive and numerous rural farm population in southern Ontario, the province has the lowest proportion of cities claiming central place functions as their dominant activity.

PRAIRIE PROVINCES

The distinctive environment of the continental interior beyond the reach of the Great Lakes-St. Lawrence waterway provides the backdrop for a distinctive city system. Focusing on the natural resources of the Canadian interior, interaction in this inland empire is facilitated by well-developed transportation networks which overcome the great distances among cities. Continentality and higher latitude has produced a climate characterized by winters with less snowfall but which are a little longer, colder and drier than those in the eastern cities; summers are warm but shorter. Relative humidity is lower, but the hours of sunshine are greater. Most cities occupy relatively flat sites adjacent to river. At an earlier period of development, Winnipeg was the primate prairie city, but the post World War II oil boom caused growth in Calgary and Edmonton which produced an urban hierarchy topped by three similar-sized cities. Reminiscent of the Atlantic Provinces, each major prairie city has its own trading area defended against competitors on the basis of sheer distance, rather than by water barriers as in Atlantic Canada.

By eastern standards the cities appear very young and uniform in layout. The uniformity of street grids is emphasized in some cities by the numbering of avenues and streets according to cardinal directions. The youthfulness of the cities and their relatively rapid growth in recent years has minimized the extent of old areas to produce cities which appear to have been completely built during the lifetimes of many of the inhabitants. Residential areas differ in appearance from those of Ontario and Quebec cities because of the widespread use of wood instead of brick. The cities are more compact and appear to have been more carefully planned. Fully serviced residential areas on the edges of cities usually end abruptly in contact with cultivated land and are not characterized by the urban sprawl that is common elsewhere in Canada.

Although the cities are younger than those of Ontario, urban populations of the two regions are similar in age structure and family size (Table 1). The substantially higher prairie birth rate, however, is second only to that of the Atlantic cities. Educational levels are slightly higher than in Ontario and Atlantic cities, chiefly on the basis of more university-trained people in the population. Prairie cities are noteworthy by having the least proportion British of all groups (except Quebec) and the highest index of ethnic diversity (Table 2). The low proportion of population which is French in origin corresponds in magnitude only to that of British Columbia. The proportion of non-British and non-French mother tongues, however, is the highest in Canada. Prairie cities differ from those of the East by the low proportion of Roman Catholics in the population and the high index of religious diversity. The portion of population born in the province of residence is lower than in eastern regions and the mobility rate is higher (Table 3).

Housing characteristics are similar to those of Ontario, through family incomes, house values and average rents are significantly lower (Table 4).

The incidence of tenancy, higher than in Ontario, is comparable to that of cities in Atlantic Canada.

The economic base of Prairie cities is quite distinctive (Table 5) and approximates that of Christaller's central places organized according to the marketing principle. The dominant functions of Prairie cities comprise a high percentage of services and retail trade (totalling fifty-six per cent), but a low percentage of manufacturing.

BRITISH COLUMBIA

The urban system of British Columbia is oriented westward toward the Pacific, rather than eastward toward the interior. More than half of the cities, including the two largest are situated on, or near, the coast; transportation networks focus strongly on metropolitan Vancouver. The urban hierarchy resembles that of Quebec: A dominant commercial metropolis, Vancouver; a government city, Victoria, about one-fifth the size of the metropolis; and a group of smaller regional cities. Tree-covered hills and mountains surround most of the cities. Tree-covered hills and mountains surround most of the cities, in contrast with the flat, agricultural areas surrounding prairie cities. Sloping topography and coastal irregularity have eliminated uniformity in street grid systems, although most cities conform to grid patterns to some extent. Urban development is not compact but sprawls instead to incorporate surrounding hill view sites. Climates are highly variable, owing to the mountainous terrain, but the majority of cities have Pacific coastal locations and enjoy moderate marine climates that are considerably milder in winter and in summer than those of inland areas. Seasons change gradually, not abruptly as in many interior regions. Because British Columbia cities, like those on the Prairies, are relatively young and the product of fairly rapid growth, they do not have extensive old areas. Houses are mainly built of wood, and stucco exteriors are more popular than elsewhere in Canada.

British Columbia cities are distinctive in a number of population characteristics. Their populations are younger than those of Prairie and Ontario cities, but older than those of cities in Quebec and Atlantic Canada (Life Cycle Index, Table 1). The high ratio of males per 100 females suggests an urban system that is oriented toward primary resource manufacturing and transportation activities in which the male labour force dominates. Educational achievement is considerably higher in British Columbia cities than in those of any other regional group--ten per cent more of the people in the cities have attended secondary school, and two per cent more have attended university, than in any other region. British Columbia cities like those of the Prairies are ethnically diverse, but are a little more British in character than those on the Prairies (Table 2). Religious diversity also is high in British Columbia cities, and they are the most Protestant of all groups. The people of British Columbia cities also are the most cosmopolitan and most mobile of all (Table 3). Compared with the other four groups, a lower percentage of the population was born within the province and higher percentages were born in other provinces and outside Canada. Both in-migration and out-migration ratios are substantially higher than those of other Canadian city groups.

The distinctiveness of British Columbia cities in housing characteristics is marked (Table 4). Single family dwellings are more common than elsewhere; apartments constitute less than twenty percent of the dwellings, some five percent less than in the second lowest region, the Prairies. House values and average rents are higher than elsewhere, but so is home ownership. On the other hand, average family income is slightly lower than in the leading city group, Ontario.

The surprisingly high percentage of British Columbia cities with manufacturing as their dominant economic function (Table 5) is caused by primary manufacturing or natural-resource processing, not by the secondary manufacturing typical of many cities in Ontario. The other prominent activity is transportation (twenty-one per cent), an industry comprising a significant percentage in the cities of only one other group, Atlantic Canada (fifteen per cent).

CONCLUSIONS

Although urban geography in Canada dates from the late 1940s, and the appearance of Taylor's urban text and several individual city studies, the comparative study of Canadian cities did not begin until the mid-1960s. Most studies were concerned with certain aspects of the national urban system and not with regionalization. Ray's study stands out as the one that advanced farthest toward a regional characterization of cities (Ray, 1971). The present study differs from that of Ray by its regional approach to cities. The limited evidence presented indicates clearly that all five city groups are distinctive, like the regions in which they are situated. As more evidence is assembled for quantitative analysis, a more comprehensive characterization of regional urban distinctiveness will be developed. The basis of such future work should comprise statistical information from many sources, including the 1981 Census of Canada.

ACKNOWLEDGMENT

Support by a research grant from the University of Victoria, and assistance from Diana Hocking, who compiled the statistical information, are gratefully acknowledged.

REFERENCES

BOURNE, L.S. and MACKINNON, R.D. (editors) (1972), *Urban Systems Development in Central Canada* (Toronto: University of Toronto, Department of Geography).

BOURNE, L.S., MACKINNON, R.D., Siegel, J. and Simmons, J.W. (editors) (1974), *Urban Futures for Central Canada: Perspectives on Forecasting Urban Growth and Form* (Toronto: University of Toronto Press).

FORWARD, C.N. (1972), Cities: Function, form and future, in A.G. Macpherson (editor), *The Atlantic Provinces* (Toronto: University of Toronto Press, Studies in Canadian Geography): 137-76.

FORWARD, C.N. (1982), The importance of nonemployment sources of income in Canadian metropolitan areas, *The Professional Geographer*, 34: 289-96.

GENTILCORE, L. (editor) (1972), *Ontario* (Toronto: University of Toronto Press, Studies in Canadian Geography).

GRENIER, F. (editor) (1972), *Quebec* (Toronto: University of Toronto Press, Studies in Canadian Geography).

HARDWICK, W.G. (1972), The Georgia Strait urban region, in J.L. Robinson (editor), *British Columbia* (Toronto: University of Toronto Press Studies in Canadian Geography): 119-33.

KARIEL, H.G. and ROSENVALL, L.A. (1978), Circulation of newspaper news within Canada, *The Canadian Geographer*, 22: 311-28.

KARIEL, H.G. and WELLING,S.L. (1977), A nodal structure for a set of Canadian cities using graph theory and newspaper datelines, *The Canadian Geographer*, 21: 148-63.

KERR, D. (1965), Some aspects of the geography of finance in Canada, *The Canadian Geographer*, 9: 175-92.

KING, L.J. (1966), Cross-sectional analyses of Canadian urban dimensions, 1951 and 1961, *The Canadian Geographer*, 10: 205-24.

LI, S-Y. (1976), Labour force statistics and a functional classification of Canadian cities, in D.M. Ray (editor), *Canadian Urban Trends, Vol. 1: National Perspective* (Toronto: Copp Clark): 58-102.

LI, S-Y., SCORRAR, D.A. and WILLIAMS, M.H. (1978), City functions, manufacturing activity, and the urban hierarchy, in R.M. Irving (editor), *Readings in Canadian Geography* (Toronto: Holt, Rinehart and Winston, third edition): 89-102.

McCANN, L.D. (editor) (1982), *Heartland and Hinterland* (Scarborough: Prentice-Hall).

MacPHERSON, A.G. (editor) (1972), *The Atlantic Provinces* (Toronto: University of Toronto Press, Studies in Canadian Geography).

MARSHALL, J.U. (1972), The urban network, in L. Gentilcore (editor), *Ontario* (Toronto: University of Toronto Press, Studies in Canadian Geography): 64-82.

MARSHALL, J.U. (1975), City size, economic diversity and functional types: The Canadian case, *Economic Geography*, 51: 37-49.

MARSHALL, J.U. (1981), Industrial diversification in the Canadian urban system, *The Canadian Geographer*, 25: 316-32.

MAXWELL, J.W. (1965), The functional structure of Canadian cities: A classification, *Geographical Bulletin*, 7: 79-104.

MAXWELL, J.W., GREIG, J.A. and MEYER, H.G. (1972), The functional structure of Canadian cities: A classification of cities, in R.M. Irving (editor), *Readings in Canadian Geography* (Toronto: Holt, Rinehart and Winston, second edition), 146-67.

MAYER, H.M. and KOHN, C.F. (1959), *Readings in Urban Geography* (Chicago: University of Chicago Press).

MERCER, J. (1979), On continentalism, distinctiveness, and comparative urban geography: Canadian and American cities, *The Canadian Geographer,* 23: 119-39.

PRESTON, R.E. (1978), The evolution of urban Canada: The post-1867 period, in R.M. Irving (editor), *Readings in Canadian Geography* (Toronto: Holt, Rinehart and Winston, third edition), 19-46.

RAY, D.M. (1971), *Dimensions of Canadian Regionalism* (Ottawa: Energy, Mines and Resources, Geographical Paper No. 49).

RAY, D.M. (editor) (1976), *Canadian Urban Trends, Vol. 1: National Perspective* (Toronto: Copp Clark).

ROBINSON, J.L. (editor) (1972), *British Columbia* (Toronto: University of Toronto Press, Studies in Canadian Geography).

ROBINSON, J.L. (1983), *Concepts and Themes in the Regional Geography of Canada* (Vancouver: Talonbooks).

SEMPLE, R.K. and Smith, W.R. (1981), Metropolitan dominance and foreign ownership in the Canadian urban system, *The Canadian Geographer,* 25: 4-26.

SIMMONS, J.W. (1967), Urban geography in Canada, *The Canadian Geographer,* 11: 341-56.

SIMMONS, J.W. (1976a), Canada as an urban system: A conceptual framework, *Ekistics,* 41: 68-75.

SIMMONS, J.W. (1976b), Short-term income growth in the Canadian urban system, *The Canadian Geographer,* 20: 419-31.

SIMMONS, J.W. (1977), *Migration in the Canadian Urban System* (Toronto: University of Toronto, Centre for Urban and Community Studies, Research Paper No. 85).

SIMMONS, J. and SIMMONS, R. (1974), *Urban Canada* (Toronto: Copp Clark, second edition).

SMITH, P.J. (1972a), Changing forms and patterns in the cities, in P.J. Smith (editor), *The Prairie Provinces* (Toronto: University of Toronto Press, Studies in Canadian Geography), 99-117.

SMITH, P.J. (editor) (1972b), *The Prairie Provinces* (Toronto: University of Toronto Press, Studies in Canadian Geography).

TAYLOR, G. (1949), *Urban Geography* (London: Methuen).

ULLMAN, E.L. and DACEY, M.F. (1960), The minimum requirements approach to the urban economic base, *Papers and Proceedings of the Regional Science Association*, 6: 175-94.

WARKENTIN, J. (editor), (1968), *Canada: A Geographical Interpretation* (Toronto: Methuen).

WONDERS, W.C. (editor) (1972), *The North* (Toronto: University of Toronto Press, Studies in Canadian Geography).

Geography Faculty, U.B.C., 1971-72

CONTRIBUTORS

Dr. Brenton M. Barr, Professor of Geography, The University of Calgary

Dr. Robert M. Bone, Professor of Geography, The University of Saskatchewan

Dr. Charles N. Forward, Professor of Geography, The University of Victoria

Dr. Hugh J. Gayler, Associate Professor of Geography, Brock University

Dr. Edward M. Gibson, Associate Professor of Geography, Simon Fraser University

Dr. Robert Gilbert, Associate Professor of Geography, Queen's University, Kingston, Ontario

Dr. Ann McAfee, Planner, City of Vancouver

Dr. Bruce Mitchell, Professor of Geography, University of Waterloo

Dr. Keith Nicol, Assistant Professor of Geography, Memorial University of Newfoundland, Sir Wilfred Grenfell College, Corner Brook, Newfoundland

Dr. Ming-ko Woo, Professor of Geography, McMaster University

Dr. James W. Young, Associate Professor of Geography, Concordia University